green guide

SEA FISHES

·································

OF AUSTRALIA

Nigel Marsh

First published in 2019 by Reed New Holland Publishers
Sydney

Level 1, 178 Fox Valley Road, Wahroonga, NSW 2076, Australia

newhollandpublishers.com

A record of this book is held at the British Library and the National
Library of Australia.

ISBN 9781925546385

Managing Director: Fiona Schultz
Publisher: Simon Papps
Editor: Liz Hardy
Designer: Yolanda La Gorcé
Production Director: Arlene Gippert
Printed in China

10 9 8 7 6 5 4

Keep up with New Holland Publishers:

 NewHollandPublishers

 @newhollandpublishers

CONTENTS

Gulf Gurnard Perch *(Neosebastes bougainvillii).*

White's Seahorse *(Hippocampus whitei).*

Bluespotted Fantail Ray *(Taeniura lymma).*

INTRODUCTION TO
AUSTRALIAN SEA FISHES

The tropical waters of Australia are home to a great variety of pretty butterflyfish including the Citron Butterflyfish *(Chaetodon citrinellus)*.

Australian waters are home to a wide variety of shark species, including the Ornate Wobbegong *(Orectolobus ornatus)*.

Australia, with more than 70,000 kilometres of coastline and a marine economic zone of more than eight million square kilometres, is home to a rich variety of sea fishes. From its northern tropical waters to its cooler temperate seas the country hosts more than 4,000 species of marine fishes, many of which are only found Down Under.

Fish have evolved to inhabit every marine environment around Australia. They are found in shallow mangroves and brackish waters, they thrive on coral reefs, rocky reefs and even sandy bottoms, and many marine fish also live in the cold, dark waters of the deep ocean. Fish are an important part of a healthy marine ecosystem, and when overfished their removal has a major impact on the health of reefs, seagrasses, mangroves and also other marine species.

This guide is an introduction to many common, and also some very strange, sea fishes found around Australia. Designed for the diver, snorkeller and angler, this guide looks at typical fish families and species, reproduction, behaviour, dangerous fishes and what we can do to protect these wonderful creatures of the deep.

Moses' Snapper *(Lutjanus russellii)*, like many fishes, are always found in schools.

The Complex Diversity of Fishes

Australia is home to a great diversity of fishes, and isolated from the rest of the world it also has many endemic species found nowhere else. Around Australia divers, snorkellers and anglers can encounter marine fishes ranging from tiny gobies to giant whale sharks.

Fish are vertebrates, the ancestors of all vertebrates, and have been around for a very long time. The first fishes evolved more than 500 million years ago, and over time they have changed and adapted into the very diverse group of animals we see today.

All fishes have similar basic characteristics, but the name applies to five classes of different animals, in the phylum Chordata, that evolved on separate evolutionary lines. Fishes all have fins for movement, gills for breathing, a supporting skeleton, a dorsal hollow nerve cord and a tail. However, they vary so much in their shape and size that identifying individual species can be a complex and difficult task. To make identification easier scientists have split fishes into classes, subclasses and families, grouping individuals with similar features.

The sharks, rays and their close relatives the chimaeras (not including in this book) have a skeleton made of cartilage instead of bone. These creatures are in the class Chondrichthyes, the cartilaginous fishes, with the sharks and rays in the subclass Elasmobranchii or strap-gilled fishes, as these creatures have five to seven gill slits. While often considered to be primitive animals because of their cartilage skeletons, sharks and rays are more highly evolved than most fishes and their reproductive behaviour is closer to mammals, breeding by internal fertilization.

The bony fishes are in the class Osteichthyes. Two subclasses exist within this group, the Actinopterygii, which contains all the well-known ray-finned fishes, and the Sarcopterygii, which contains the more primitive fleshy finned fishes, such as the lungfish. The final class of living fishes (there are also two classes of extinct fishes) is the Agnatha or jawless fishes (not included in this book).

Fish are found in all marine and freshwater aquatic environments, and with more than 500 families and 33,600 fish species so far described by science, this group of animals is the most complex and diverse of all the vertebrates.

Fishes vary greatly, some have scales like the Black-spot Goatfish *(Parupeneus spilurus),* while others have tough leathery skin like the Eastern Smooth Boxfish *(Anoplocapros inermis).*

Sharks also vary greatly in their shape and size, with the Tasselled Wobbegong *(Eucrossorhinus dasypogon)* having a flattened camouflaged body to aid in ambushing prey.

Many bottom-dwelling fish have striped or banded patterns to help with camouflage, such as the Magpie Morwong *(Cheilodactylus gibbosus)* and Moonlighter *(Tilodon sexfasciatus)*.

Typical Fish Features

A shark, stingray and a clownfish may look very different, but like all fishes they have similar body features.

Fins

All fish have fins. They can be large, small, decorative, or more like hands, but they all serve the same function of locomotion. The tail or caudal fin is used to drive a fish forward, the pectoral fins are used for turning, while the dorsal, ventral and anal fins are for stability. However, in some fishes the fins have evolved for different uses. In suckerfish the dorsal fin has changed into a suction cup to allow these animals to stick

Many bottom-dwelling fish like the Whitestreak Grubfish *(Parapercis stricticeps)* use their fins to perch on the bottom.

to other fish. In clingfish the pelvic fins have also changed into a suction disk so these fish can cling to rocks, corals and seaweeds. In rays the pectoral fins have fused to the body to create a large body disc. While lionfish have long feather-like pectoral and dorsal fins that they fan out to round up small fish.

The Giant Frogfish *(Antennarius commerson)* uses its fin-like hands and feet to walk across the bottom.

Gills

To breath underwater all fish have gills that extract oxygen from water. Bony fish have one set of gills on each side of the head, while sharks have five to seven gill slits on each side of the head and rays have five gill slits on their ventral surface. Most fish, sharks and rays suck water through their mouth and out through their gill slits, passing over the gills in the process. However, a number of ray species, which bury in the sand, can also inhale water through a hole behind the eye, called a spiracle. This allows them to remain covered and breathe at the same time.

Sharks, like this Greater Bamboo Shark *(Chiloscyllium magnus)* typically have five gill slits.

Scales, Skin and Slime

Most bony fishes have skin covered in scales. These scales vary greatly in size, shape and strength and can be used to identify fish species. Scales grow out of the skin, similar to hair in other animals, and overlap like roof tiles to form an interlocking barrier. Scales form a shield to protect fish against parasites, predators and diseases, and are generally covered in slimy mucus. Scales also serve to make fishes more hydro-dynamically efficient through the water. Sharks and rays have very different scales which are more like modified teeth, giving them sandpaper-like skin. Not all fish have scales, as some have leather-like skin that they moult regularly, others have soft skin covered in smelly mucus and some have a tough exoskeleton.

The Lunartail Bigeye *(Priacanthus hamrur)* has large eyes as it is a nocturnal hunter.

Senses

Fishes have well-developed brains and keen senses, and contrary to claims, they do feel pain and don't have short memories. Vision is a very important sense for most fishes, and used to find food and avoid predators. Many nocturnal fishes have eyes like a cat that allow them to see very well in the dark. Hearing is another important sense used by fishes, especially with water 800 times denser than air. Fish can also detect sounds and vibration through their lateral line, a series of modified hair cells that runs down each side of the body and are also used to sense movement and pressure changes. Some fish can also perceive sounds through their swim bladder, an organ used to control buoyancy.

Fishes have a good sense of smell and taste through the use of chemoreceptors.

These receptors are located in the mouth and nostrils, but some fish also have chin barbels with receptors and other fish have receptors all over their bodies. Fishes also have additional senses that other animals don't possess. They can detect pressure differences, water movement, electric signals from other animals and also the earth's magnetic fields.

Body Shapes

The body shape of fishes varies dramatically, and is the best feature to identify the family or the species. Fish bodies can be elongated, short, squat, flat, oval, circular, box-like, triangular and even snake-like. This vast difference in body shapes is mostly driven by the habitat the fish lives in. Fishes that live in holes and caves have elongated bodies that allow them to squirm into tight recesses. Fishes that spend a lot of time on the bottom have small fins as they don't swim much, while fishes that constantly swim around are streamlined with powerful tails.

Dangerous Australian Sea Fishes

While most people think sharks are the most dangerous creatures in the oceans, there are many other fishes that are potentially more deadly.

Many fishes are highly venomous, and a number of people have died from being spiked by one of these toxic creatures. Fish have venomous spines for defence against predators. Most fish venom, such as from the catfish, rabbitfish and rays, has minor effects on humans, mainly causing pain and discomfort. But members of the scorpionfish family, which includes the stonefish, are potentially deadly. These camouflaged fish are sometimes touched or stepped on by accident, and the venom is fast acting and causes agonising pain. If ever stung by any scorpionfish, immerse

Harmless looking reef fish, like the Sixspine Leatherjacket *(Meuschenia freycineti)*, often have defences that make them potentially dangerous, in the case of leatherjackets it is a sharp dorsal spine.

the wound in very hot water and seek immediate medical assistance.

A small number of shark species are potentially dangerous. However, most shark bites are an accident on the shark's part, mistaking a human for potential prey. In reality, most sharks are small and docile and only a handful are potentially dangerous.

The Monotone Moray *(Gymnothorax monochrous),* like all moray eels, has very sharp teeth and should always be treated with respect.

Stingrays use their tail barbs for defence against sharks, but will also turn them on people when threatened.

There are many other fishes that sometimes bite divers, snorkellers and swimmers, including moray eels, triggerfish and damselfish. But attacks from these fishes are very rare and most injuries are minor. Other fishes like flatheads, surgeonfish and leatherjackets are potentially dangerous to anglers, as they have sharp spines.

Many fishes found in Australian waters are poisonous and should never be eaten. Fishes with toxic skin or flesh include the puffers, porcupinefish, moray eels and some boxfish. In tropical reef waters many fishes have toxic flesh from a build-up of microscopic dinoflagellates, a type of algae. If eaten it leads to ciguatera poisoning. Fishes that carry ciguatera include the triggerfish, rockcods and gropers. Fishes at the top end of the food chain, like sharks and gropers, can also have high levels of mercury and should never be eaten.

The well camouflaged Estuarine Stonefish *(Synanceia horrida)* – one of the most dangerous fish species found in Australia.

Deceptive Fishes

To avoid being eaten, fishes have developed a range of tactics to remain alive. Some

The Common Cleaner Wrasse is one of the most popular fish on the reef due to its cleaning habits.

fish have toxic skin or flesh to make them unpalatable, while other fish use concealment or camouflage to keep them safe. However, there are some fish that use deception to avoid becoming a meal.

The most common form of deception is mimicry. Fish mimics have developed a body shape and colouration that matches other fish that are either poisonous or venomous. The best known of these is the Mimic Filefish (*Paraluteres prionurus*) which is a type of leatherjacket that looks exactly like the Saddled Puffer (*Canthigaster valentini*). Another fish that deploys mimicry to keep it safe is the False Scorpionfish Rockcod (*Centrogenys vaigiensis*). This tropical species sits on the bottom and fans its fins so it looks like a venomous scorpionfish.

While most deceptive fish mimic other fish, a number also mimic toxic flatworms and sea slugs to keep them safe. This tactic is most often used by juvenile fish, especially bottom-

The False Cleanerfish looks almost identical to the Common Cleaner Wrasse and uses this deception to sneak up and nip the flesh off other fish.

dwelling frogfish. Juvenile Warty Frogfish (*Antennarius maculatus*) and Painted Frogfish (*Antennarius pictus*) have colour patterns that are similar to flatworms and nudibranchs. A number of soles also mimic flatworms, not only having a similar colour pattern, but moving in the same fashion as the flatworm.

Mimicry is not just used to keep fish safe from predators, as some use it to obtain

a meal. The False Cleanerfish (*Aspidontus taeniatus*) is a type of blenny that mimics one of the best-loved fish on the reef, the Common Cleaner Wrasse (*Labroides dimidiatus*). It has a similar body shape and colouration, and even swims like a Common Cleaner Wrasse. It uses this deception to get close to fish who think they are about to get a clean, but instead the False Cleanerfish nips a mouthful of flesh and retreats to its hole.

The juvenile Warty Frogfish has a colour pattern that is very similar to a toxic nudibranch.

The Saddled Puffer, like all pufferfish, has poisonous flesh.

The Mimic Filefish has developed a colour pattern similar to the Saddled Puffer to keep it safe from predators.

The False Scorpionfish Rockcod sits on the bottom pretending to be a venomous scorpionfish.

Fish Chameleons

A surprising number of fish can change their colour, just like chameleons. These colour changes are a result of stress, age, reproduction, gender and camouflage. Fish, like many cold-blooded creatures, can change their colour due to cells in the skin called chromatophores.

Within fish skin are three types of chromatophores; these are xanthophores, which contain yellow and red pigments, melanopores, which contain black pigment and iridophores, which are colourless crystals that reflect light to generate blues, whites and ultraviolet (these give many fishes their shiny metallic appearance). Fish colour changes are mainly controlled by the movement of black pigment in the melanopores, and by concentrating or dispersing these blacks it allows other colours to show through. Some fishes can do this almost instantly, but they are generally limited in their colour palate. This instant colour change is common for surgeonfish and trevallies that go from a light shade to a dark shade in the blink of an eye. Many bottom-dwelling fish like flounders, soles and leatherjackets can also change colour as they swim over different terrain to assist with camouflage.

The Blue Angelfish *(Pomacanthus semicirculatus)*, is a very colourful reef fish, but the juvenile (below) of this species is even more spectacular with its striking colour pattern.

Frogfish change colour to help with camouflage, but it is a process that can take a few days. Fish without scales, that moult their skin regularly as they grow, can also select a new skin colour. The best known example of this is the Leaf Scorpionfish

(*Taenianotus triacanthus*) that moults about every two weeks.

Many fish have different colours as juveniles, changing colour as they grow and get

older. Some juveniles have more camouflaged skin colours and patterns, but others are much more garish. The angelfish are a good example, with the juveniles often having much more spectacular colours than their parents. Some fish species have such different colours for males, females and juveniles that scientists once thought they were completely different species.

Many fish change colour when ready to reproduce, to show they are receptive to mating. They also change colour when they are stressed. Fish captured by anglers often lose their vivid colours when removed from the water.

Some fish change colour throughout the day, with the biggest change coming at night. Many fish sleep in

The Common Coral Trout *(Plectropomus leopardus)* is usually a pinkish-brown colour, but at night it changes to a mottled pattern for camouflage.

holes in the reef at night, and change into sleeping colours. At night their colours are more muted and dull, which is thought to aid their concealment and keep them safe from predators.

The Leaf Scorpionfish moults its skin every two weeks and can change colour at that time.

Why Do Fish School?

Many fishes form into groups, for a variety of reasons. Most fish school together as a form of protection against predators. Being one of a vast school gives fish safety in numbers, making it harder for predators to pick off individuals. They also school together as a social behaviour, making it easier to reproduce, as they don't have to go out and search for a mate. Fish also gather together in schools for migrations, moving up and down the coast with the seasons.

Many sharks and rays also school, but their dynamics are often far more complex. Rays mostly gather together in schools to feed and mate, and are quick to go their separate ways. While sharks that school together are always of the same sex, with the males and females in different groups and only coming together at certain times of the year to mate.

Roundface Batfish *(Platax teira)* are always found in schools that can sometimes number in the hundreds.

Turrum *(Carangoides fulvoguttatus),* a species of trevally, hunt in packs on smaller fish.

Fish often form into mixed schools for safety in numbers, like these Black-spot Goatfish *(Parupeneus spilurus)* and Yellowfin Bream *(Acanthopagrus australis).*

Fish Parasites

Fish, like all creatures, can get infected by parasites and other bugs. Most of these parasites are internal and never seen by people, but a few external parasites are far more obvious.

The best known of these parasites are the isopods. These bloodsuckers are a type of crustacean, with the family containing more than 10,000 species. Not all isopods are parasitic, as many are scavengers and feed on dead animals. However, the parasitic ones have a liking for flesh and blood, and permanently attach to fish. Most isopods are less than 2 cm long and generally attached to the head or body of the fish, but there is one species of isopod that likes to attach in the mouth, eating the tongue!

This Blackspot Toby *(Canthigaster bennetti)* sports a strange pair of eyebrows, clusters of copepods.

Another family of crustaceans that attach and feed off fish are the copepods. This family contains around 13,000 species that vary greatly in form and lifestyle. Most copepods are free-living and drift with ocean currents, but a number are parasitic and feed on the flesh and blood of fish and other marine animals. Copepods have needle-like mandibles and once they latch onto a fish they are almost impossible to get off. They are often found attached to the eyes, but will also attach to the mouth and gills. Both isopods and copepods can kill their host, but many fish live a full life with one of these bloodsuckers attached.

Another parasite observed on fish are leeches. These parasitic worms are more common on sharks and rays, and feed on the blood of the animal. Some leeches are so specialised in their feeding that they only attach to one species, with the Gilled Leech *(Branchellion* sp.) only found on Coffin Rays *(Hypnos monopterygius)*. There are thought to be around 100 species of marine leeches.

This isopod is attached to the head of a Cook's Cardinalfish *(Ostorhinchus cookii)*.

The Gilled Leech is only found attached to Coffin Rays.

How Do Fish Clean Themselves?

Almost all fish like a clean and they regularly visit cleaning stations to get this important service done. Lacking hands, fish cannot clean themselves of old scales, parasites or food stuck between their teeth. But several species of fish, and also a few shrimps, have stepped up to provide cleaning services.

Around 50 species of fish have been found to clean other fishes. The best known is the Common Cleaner Wrasse (*Labroides dimidiatus*). These small fish, which only grow to 12 cm in length, establish cleaning stations and fish queue up for a service.

They pick over the body of their clients and also boldly enter their gills and mouth. They also clean sharks and rays, and when a giant Reef Manta Ray (*Mobula alfredi*) arrives to be cleaned, a whole team of Common Cleaner Wrasses go to work.

A pair of White-banded Cleaner Shrimps go to work on a Titan Triggerfish (*Balistoides viridescens*).

Fish also visit cleaning stations established by a variety of shrimp species. These shrimps operate cleaning stations in caves, ledges and sometimes sea anemones. The best known of these cleaners

Common Cleaner Wrasse boldly enter the gills and mouth of fish to perform their cleaning duties.

are the White-banded Cleaner Shrimps (*Lysmata amboinensis*). These pretty, small shrimp jump onto fish when they need a clean, but sometimes their services are monopolised when a moray eel moves into the same hole.

Fish Reproduction

Fishes reproduce in a variety of different ways. The bony fishes are typically spawners, with the female and male simultaneously releasing a multitude of eggs and sperm. This spawning can happen between a couple, a group or thousands of fish at the same time. Once a sperm finds an egg and fertilisation occurs, the developing young will typically drift with ocean currents until it finds a new home.

But not all bony fishes reproduce in such a haphazard way. Many female fish lay their eggs on rocks or corals and the male then fertilises them. Some of these fish also guard the eggs until they hatch, then the young have to fend for themselves. Other bony fishes take parenting much more seriously. Male seahorses and their relatives keep their fertilised eggs in a special brooding pouch until they hatch, and other fishes stick their eggs on the side of their body until they hatch. But the most dedicated parents are the mouth brooders, fishes that keep their fertilised eggs in their mouth until they hatch.

The male Pink Anemonefish *(Amphiprion perideraion)* tenderly checks its eggs.

A Bengal Sergeant *(Abudefduf bengalensis)* tends its clutch of eggs laid on a rock.

The reproduction strategy used by bony fishes varies greatly, but their sex lives can be even more complex. Most fishes are either male or female, but a number can change sex. Some fish families, like the wrasses, are dominated by a male who breeds with a harem of smaller females. If he dies, the largest female changes sex and takes his place. Other fish families, like the anemonefish, are dominated by a large female, who has a male partner and smaller male assistants. If she dies her male partner changes sex and one of

Some of the best fed fish on the Great Barrier Reef are the famous Potato Cod *(Epinephelus tukula)* of Cod Hole.

Nurse Shark (*Carcharias taurus*). A close encounter with a shark or ray is an exciting experience, and once you have seen one of these graceful animals up close you quickly realise they are not mindless killers, but just another large fish. Only a handful of shark species are potentially dangerous, and staying out of the water because of a fear of sharks means you are missing out on some wonderful interactions with some amazing fishes.

One of the friendliest fish divers encounter in New South Wales is the cheeky Eastern Blue Groper.

Protecting Australian Fishes

Many fish species are captured for food by both commercial and recreational fishers. However, overfishing has seen a horrendous depletion of fish numbers around the planet, and Australia is little different. Protecting fishes, and their habitats, is important not only to ensure fish populations, but also the health of the oceans.

Marine Reserves

The best way to protect fishes is with a network of marine reserves, where fish and other marine life are fully protected. Around Australia are numerous marine reserves, with the largest being the Great Barrier Reef Marine Park. Marine reserves give fish a safe place to live and breed, and research over many decades has shown they are very successful. Studies have shown that fish are larger and more abundant in marine reserves, compared to nearby areas where fishing occurs. The beauty of these marine reserves is the overflow effect, as when the fish breed, their young populate other areas.

Size and Bag Limits

Fisheries around Australia set size and bag limits on popular fish species to ensure they are not overfished. While these are set with good intentions, they can still lead to the decline of some fishes. The biggest fish are the best breeders, releasing more eggs and sperm than small fish. Unfortunately many anglers take these big fish and throw back the small fish, and researchers have found that many fish are now breeding at a smaller size and producing fewer offspring. Anglers need to be concerned about the fish they keep and the ones they throw back. Many fish, especially ones pulled up from deep water, die when returned to the water, from a burst swim

Always dispose of fishing line and fishing gear responsibly and never in the ocean. Thousands of fish die each year tangled in discarded fishing line like this poor moray eel.

bladder, exhaustion or by being eaten by sharks or predatory fishes.

Sustainability

People will always eat fish; it is a very healthy food source and if managed properly is sustainable. However, to ensure it is sustainable people need to be educated about what fish are protected and what fish can be eaten. Anglers need to know fish species, so they can identify protected species and release them unharmed. While the consumer needs to know what fish are sustainable and can be safely eaten without impacting on fish populations. The Australian Marine Conservation Society publishes a wonderful Sustainable Seafood Guide (www.sustainableseafood.org. au) which can be downloaded as an app.

Overfishing for sharks has seen many species depleted and has left many species endangered.

Sharks are still sadly fished and killed for sport, with the anglers thinking they are performing a public service. However, sharks are an important part of a healthy marine ecosystem.

One of Australia's most iconic fish is the magnificent Leafy Seadragon *(Phycodurus eques)*.

AUSTRALIAN SEA FISHES

With more than 4,000 species of fish found in Australian waters this guide would be as big as a phone book if it were to include every fish family and every fish species. Instead, over the following pages are the more common and interesting fish families found in Australia via the marine habitats where they are found or by a shared behaviour trait.

Grubfish, like the Pinkbanded Grubfish
(*Parapercis nebulosa*), live and feed on sandy bottoms.

THE DESERT DWELLERS

At first glance a sandy bottom can seem devoid of fish species. However, these environments are not underwater deserts, as many fish feed on the numerous invertebrate species that live in the sand and many fish live on or hide under the sand.

Angel Sharks

Australian waters are home to a rich and varied number of sharks, with 30 families and around 160 species so far identified. Of these only a handful are considered potentially dangerous, and in reality most shark species are small, shy and completely harmless. Sharks inhabit all marine environments in Australia, and a number are found on sandy bottoms.

The best-adapted sharks for a sandy environment are the angel sharks. These flat-bodied sharks are ambush predators that like to hide under a layer of sand while waiting for prey. When a fish or crustacean gets close to their mouth the angel shark explodes from the sand and snatches its meal. Australia is home to four species of angel shark, but only the Australian Angel Shark (*Squatina australis*) is commonly seen by divers and snorkellers. Growing to a length of 1.5 m, the Australian Angel Shark is found in Australia's temperate waters.

Only found in southern waters, the Australian Angel Shark is an ambush predator.

Leopard Shark

The prettiest of all the shark species is generally found on sandy bottoms, the wonderful Leopard Shark (*Stegostoma fasciatum*). In a family all by itself, this species' closest relative is thought to be the Whale Shark. Found in Australia's tropical and subtropical waters, Leopard Sharks rest by day and stalk the sand by night, feeding on

Leopard Sharks spend most of the day resting on the sand.

fish, molluscs and crustaceans. Growing to a length of 2.5 m, they are rarely seen by divers in most areas of Australia but are fairly common off southern Queensland and northern New South Wales during the warmer months.

Electric Rays

The great majority of Australian ray species inhabit sandy bottoms, feeding and hiding in this environment. Australian waters are home to 16 families of rays and more than 100 species, and while some are potentially dangerous, most rays are docile and wary of divers.

The most unusual family of rays found on sandy bottoms are the electric rays. Split over three families, eight of these strange rays are found in Australian waters. Able to generate an electric current by rubbing

Able to generate an electric shock of 200 volts the Coffin Ray is a fish best avoided.

modified muscles, electric rays use these chargers to stun prey and for self-defence. The electric ray typically found around most of Australia is the Coffin Ray (*Hypnos monopterygium*). Growing to a length of 60 cm, these chubby rays are often found by accident, but able to generate a jolt of 200 volts it is an encounter that a diver never forgets.

Shovelnose Rays

With a head like a ray and a body like a shark, shovelnose rays are often mistaken for sharks, but like all rays they have their gills on their underside. The shovelnoses are actually split into four families and 12 of these odd-shaped rays are found in Australian waters. One of the largest members of this family is found in tropical waters, the White-spotted Wedgefish (*Rhynchobatus australiae*). A member of the Wedgefish family, this species can grow to 3 m in length. A shovelnose ray regularly seen in Australia's cooler temperate waters is the Southern Fiddler Ray (*Trygonorrhina dumerilii*). This species has a round head with a pretty banded pattern and grows to 1 m in length. Like other small shovelnose rays it likes to hide under a layer of sand.

Southern Fiddler Ray.

White-spotted Wedgefish.

Stingarees

The stingarees look like a miniature version of a stingray, but with a much shorter tail. Australia is home to the largest variety of these small rays, with around 21 species identified. Stingarees are typically found feeding and hiding on sandy bottoms, and in some areas dozens can be found resting together. The species most frequently seen off New South Wales is the aptly named Common Stingaree (*Trygonoptera testacea*). These sandy coloured rays are found in bays and reach a disc width of 30 cm. Stingarees

Common Stingaree.

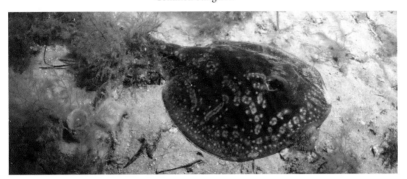

Spotted Stingaree.

are actually most abundant in cooler temperate waters, with the Spotted Stingaree (*Urolophus gigas*) seen from Victoria to southern Western Australia. These pretty rays have a black and grey blotched-circular pattern across their disc.

Stingrays

Although potentially dangerous because of their tail barbs, stingrays are in fact docile animals that would rather flee from a predator or diver. Feeding on fish, crustaceans and just about anything they find in the sand, stingrays use electro-sensors on their

One of the most common stingrays in tropical waters is the Blue-spotted Maskray.

Smooth Stingray.

Porcupine Stingray.

snout to locate buried prey. Australia is home to 22 species of stingrays and almost all are found in tropical waters.

One of the most common tropical species is the small Blue-spotted Maskray (*Neotrygon australiae*). Only growing to 40 cm wide, it is frequently seen on the Great Barrier Reef. A rarer tropical species is the Porcupine Stingray (*Urogymnus asperrimus*). This unusual stingray lacks a tail spine and is instead covered in sharp thorns. The only species generally found in southern Australia is the Smooth Stingray (*Bathytoshia brevicaudata*). These giant rays are regulars at boat ramps and jetties, cleaning up the scraps left by anglers, and can be more than 2 m wide and weigh up to 350 kg.

Skates

All rays give birth to live young, except for the skates which lay eggs. These unusual rays also lack tail barbs and instead have small thorns on their back and tail for defence. Australia is home to 25 species of skate and most are found in deep water. But two species venture into the shallows in temperate waters and are sometimes seen by divers.

Thornback Skate.

Melbourne Skate.

The most common of the pair is the Thornback Skate (*Dentiraja lemprieri*). This small ray forages in bays, digging in the sand for invertebrate species. The Melbourne Skate (*Spiniraja whitleyi*) is the largest skate species in Australia, reaching a width of 1 m. Sometimes seen in Melbourne's Port Phillip Bay, these skates prefer deep water.

Snake Eels

The eel family is vast and varied, with 19 families found in Australian waters. Many of these are found in deep water, and some in freshwater, but two families of eels inhabit sandy environments. Snake eels live in the sand and could best be described as being like a giant worm. The only part of a snake eel most divers see is the head, with the rest of the body buried in the sand. Australia is home to 45 species of snake eels with the Serpent Eel (*Ophisurus serpens*) being the species most frequently seen in temperate waters. This sandy coloured eel is easily overlooked as it is well camouflaged on sandy bottoms. But reaching a length of 2.5 m it is one of the larger species of snake eels.

Garden Eels

Garden eels are a member of the conger eel family that live in large colonies that can number in the thousands. These bizarre eels live in the sand and are usually seen

The Serpent Eel is mostly seen at night.

Spotted Garden Eels are always found in large colonies.

with their head and part of their body exposed, swaying back and forth catching zooplankton. The most common species seen in Australia is the Spotted Garden Eel (*Heteroconger hassi*). Like all garden eels it is a tropical species, and best observed on the sand flats of the Great Barrier Reef.

Catfish

Most catfish species inhabit freshwater lakes and rivers, but Australia is also home to a small number of marine species. Catfish typically have an eel-like tail and fleshy barbels around the mouth, which look like cats' whiskers, which they use to find food. Catfish are generally found on sandy bottoms searching for their favourite food of small invertebrates.

Estuary Catfish.

The most common tropical catfish species is the Striped Catfish (*Plotosus lineatus*). These black-and-white striped catfish grow to a length of 35 cm and are found in schools. When feeding they constantly leapfrog each other as they search for morsels of food in the sand. A larger catfish species found in cooler waters from southern Queensland to southern Western Australia is the Estuary Catfish (*Cnidoglanis macrocephalus*). Also called the Estuary Cobbler, this species grows to 60 cm in length and is usually found in estuaries.

Striped Catfish.

Gurnards

Gurnards are strange fish with armour-plated heads and wing-like pectoral fins. These unusual fish feed on sandy bottoms, searching for crustaceans, cephalopods and

The Spiny Gurnard varies in colour depending on the region.

small fish to consume. More frequent in deep water, Australia is home to 33 species of gurnards, but only a handful of species are seen by divers.

The most common and widespread member of this family is the Spiny Gurnard (*Lepidotrigla papilio*). This species is found in temperate waters around southern Australia and grows to 20 cm in length. Spiny Gurnards vary in colour depending on their location. Off Victoria they are a creamy colour with light blue wings, while off New South Wales they are bright red with vivid blue wings. When threatened Spiny Gurnards fan out their pectoral fins and flash their dazzling colours to confuse predators.

Flatheads

A popular eating fish, flatheads are well represented in Australian waters with 45 species identified. These flat-bodied fish are mostly found on sandy bottoms, spending much of their time hidden under a layer of sand. Flatheads prey on a variety of invertebrates, small fish and cephalopods. Tropical flatheads tend to be small in size and nocturnal, while temperate species are larger and targeted by anglers.

One of the largest tropical species is the Northern Rock Flathead *(Cymbacephalus staigeri),* which reaches a length of 50 cm. A well-camouflaged fish, they rest on both sand and coral reefs, relying on their mottled

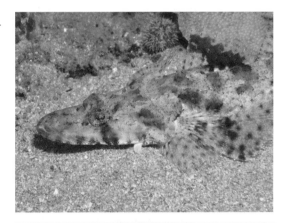

The Northern Rock Flathead (above) is a tropical flathead, while the Dusky Flathead (below) is common off New South Wales.

skin patterns to keep them hidden. The largest member of this family is the Dusky Flathead (*Platycephalus fuscus*). This species is regularly seen in New South Wales and southern Queensland and grows to 1.2 m in length.

Whiting

Popular with anglers and commercial fishers, whiting are elongated fish that are often found in schools on sandy bottoms. These fish feed by digging in the sand, and have a well-developed chemosensory system on their snout. Whiting feed on a range of invertebrate species, including crustaceans, molluscs and even echinoderms. In Australian waters there are 13 species of whiting.

Sand Whiting feed on sandy bottoms.

The Sand Whiting (*Sillago ciliata*) is most frequently captured by anglers. Found along the east coast of Australia, it is common in bays, estuaries and surf beaches. This species is a silver-grey colour with yellow ventral and anal fins, and grows to 50 cm long.

Goatfish

Goatfish are easily identified by their long chin barbels, which they use to locate food when digging in the sand. Mostly found in tropical waters, Australia is home to 26 species of goatfish. Goatfish are usually seen in small groups, but some species also form into large

Black-spot Goatfish.

schools. Able to change colour, goatfish are often a pale colour when on sand and more brightly coloured when resting on corals. A few species also mimic the colour patterns of other fish. This allows them to mix with these fish and receive protection by hiding within the school structure.

The Black-spot Goatfish (*Parupeneus spilurus*) is a widespread member of this family, common in subtropical and warm-temperate seas. Like many goatfish it has a striped pattern, but with a distinctive black spot before the

Southern Goatfish.

tail. One of the largest of all the goatfish, this species can grow to 50 cm in length. The only member of this family regularly seen in temperate waters is the Southern Goatfish (*Upeneichthys vlamingii*). This species varies greatly in colour depending on the size of the fish and the time of day. Generally they are a reddish-yellow colour with many blue spots and lines. Growing to a length of 40 cm, the Southern Goatfish is generally found in sheltered bays.

Grubfish

Grubbing in the sand for food, grubfish are a very appropriately named family of fish. These elongated fish are often very curious of divers, and will closely follow the fins of a diver in the hope that they will kick up some prey for them to snack on. Found in both tropical and temperate waters, Australia is home to 28 species of grubfish.

Most grubfish have pretty skin patterns, and the Blacktail Grubfish (*Parapercis queenslandica*) is no exception. This tropical species grows to 25 cm in length and has a distinctive black and white speckled pattern and a characteristic black blotch on the tail. Grubfish are less common in Australia's temperate waters, but one species to

Two common Australian grubfish are the Blacktail Grubfish (above) and the Spotted Grubfish (below).

43

look out for is the Spotted Grubfish (*Parapercis ramsayi*). This species has a fine spotted pattern on its back and large spots on its stomach. It is thought to be more common in deeper water, but divers encounter Spotted Grubfish in New South Wales, South Australia and southern Western Australia.

Stargazers

Stargazers are rarely seen by most divers as they spend much of their time buried in the sand with only their mouth and eyes exposed. These strange fish are ambush predators, snatching prey when it passes close to their high-set mouth. Stargazers also have venomous spines

The Common Stargazer is rarely seen as they hide under a layer of sand.

behind the head for defence and a number of species can also generate an electric shock. Australia is home to 20 species of stargazers, but of these the Common Stargazer *(Kathetostoma laeve)* is the most widespread and abundant. Growing to a length of 75 cm, the Common Stargazer is best observed at night, when they sometimes relocate to a new hiding spot.

Gobies

With more than 2,000 species, the gobies are the largest fish family. While found in a variety of marine environments, gobies are usually found on sandy bottoms, with many species living in burrows. Gobies are

Black-lined Sleeper Goby.

small bottom-dwelling fish that feed on tiny invertebrates. Australia is home to 355 goby species, which are mostly found in tropical waters.

One of the most widely distributed gobies in Australian waters is the Whitebarred Goby (*Amblygobius phalaena*). This goby is found in tropical and subtropical waters, and like many gobies it is always found in a male and female pair. The Whitebarred Goby is quite a large member of the family, reaching a length of 15 cm. A smaller tropical species often seen on the Great Barrier Reef is the Red-lined Goby

(*Amblygobius rainfordi*). This species has pretty red stripes and grows to 6.5 cm long.

Sleeper gobies are generally larger than their cousins and have vivid markings and a metallic sheen. The Black-lined Sleeper Goby (*Valenciennea helsdingenii*) is one of the more common members of this group and found along the east coast of Australia. This species is easily distinguished by the two lines that run the length of its body, which can vary in colour from black to orange. Black-lined Sleeper Gobies grow to 25 cm in length and are often observed gulping mouthfuls of sand to sieve for food.

One of the strangest groups of gobies found in sandy environments are the shrimpgobies. These small fish share a hole with shrimps, and in a symbiotic relationship the shrimps maintain the home while the gobies watch for

Two common reef gobies are the Orange-spotted Shrimp Goby (left) and the Red-lined Goby (right).

predators. A number of these gobies are found in the Australian waters, including the Orange-spotted Shrimp Goby (*Amblyeleotris guttata*). This species is often seen on the Great Barrier Reef and reaches a length of 11 cm.

Gobies are not only found on the sand as many species make a home on corals. The Seawhip Goby (*Bryaninops yongei*) is one of a number of gobies that like to live on seawhip corals. These small fish only reach 4 cm in length and with a semi-

The Seawhip Goby lives its entire life on a sea whip.

The Whitebarred Goby is quick to disappear into its hole when disturbed.

transparent body they are easily overlooked. The Seawhip Goby is a tropical species and common on the Great Barrier Reef.

Flounders

The flounders are flat-bodied fish that spend much of their time hidden under a layer of sand. They are more active at night when they hunt the sand for invertebrates and small fish. This complex family contains a number of branches, including the sand flounders, largescale flounders, lefteyed flounders and righteyed flounders. In total around 82 species of flounders are found in Australian waters.

Greenback Flounder.

Flounders are found in both tropical and temperate waters around Australia and many are targeted by anglers. The Smalltooth Flounder (*Pseudorhombus jenynsii*) is a member of the sand flounders and found in temperate waters around southern Australia. This species grows to 35 cm in length and has a spotted pattern. The tropical Leopard Flounder (*Bothus pantherinus*) looks very similar to

Smalltooth Flounder.

this species and their ranges overlap in New South Wales. One of the larger members of this family is the Greenback Flounder (*Rhombosolea tapirina*). A member of the righteyed flounders, this species is almost diamond shaped and reaches a length of 45 cm. The Greenback Flounder is found in temperate waters throughout southern Australia, often in shallow bays.

Flounders, and the closely related soles, start life with their eyes on either side of their body like other fish. Juvenile flounders and soles live a pelagic life until they are ready to settle on the bottom. Once this occurs one of their eyes migrates to join the other eye on one side of the head. In flounders this can be either side of the body depending on the species, while in soles it is always the right side.

Soles

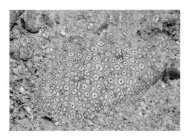

Peacock Sole.

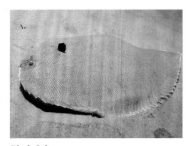

Black Sole.

Soles look very similar to flounders but are generally much smaller in size and also have smaller eyes and mouth. Many also have much prettier patterns than the flounders, and some even mimic toxic flatworms in their colouration and movement to keep them safe from predators. Soles are more common in tropical waters and around 59 species are found in Australian waters.

One member of this family often found in tropical waters is the Peacock Sole (*Pardachirus pavoninus*). This species is a sandy colour with numerous blotches. The Peacock Sole grows to 22 cm in length and has a closely related cousin found off New South Wales called the Southern Peacock Sole (*Pardachirus hedleyi*). The largest sole found in Australia is the Black Sole (*Brachirus nigra*). This species grows to 35 cm in length and is popular with anglers. Found from southern Queensland to Victoria, the Black Sole is generally a sandy colour with a small black pectoral fin.

Hiding in the seagrass, the Mossback Velvetfish *(Paraploactis trachyderma)* is a very cryptic fish.

THE CRYPTIC AND CAMOUFLAGED

The following fish families are sometimes difficult to find as they like to stay hidden. Some have camouflaged skin patterns that help them to blend in with the bottom, while others are shy and use cryptic skills to keep them concealed. These camouflaged and cryptic fishes use these talents to either ambush prey or to avoid being eaten themselves.

Wobbegongs

Australia is home to many cryptic shark species, with the best known of these being the wobbegongs. Ambush predators, wobbegongs have camouflaged skin patterns and dermal lobes around the head, which look like fleshy beards and help to conceal their presence. Wobbegongs feed on a wide variety of marine life and have very sharp teeth, so should never be touched or harassed.

Banded Wobbegong.

Australia is home to the world's largest variety of wobbegongs, with ten species found around the nation. One of the most abundant and widespread species is the Spotted Wobbegong (*Orectolobus maculatus*). The largest of all the wobbegongs, growing to 3.2 m in length, the Spotted Wobbegong is found in temperate waters from southern Western Australia to southern Queensland, but is more frequently seen off New South Wales. Also found in the same range is the Banded Wobbegong (*Orectolobus halei*), which is also quite large, reaching 2.9 m in length, but is a shy species that often shelters in caves.

Northern Wobbegong.

Spotted Wobbegong.

A number of wobbegongs are also found in Australia's tropical waters, including the rarely seen Northern Wobbegong (*Orectolobus wardi*). This species only reaches 1 m in length and is best found off northern Western Australia. Divers sometimes see this species while exploring Ningaloo Reef.

Catsharks

The catshark family is very large and well represented in Australian waters, with 34 species recorded. Shy and reclusive, most catsharks are nocturnal and spend the day hidden in caves, or live in deep water, so are rarely seen. The only catshark species

The Draughtboard Shark likes to hide in kelp forests.

regularly seen by divers is the Draughtboard Shark *(Cephaloscyllium laticeps)*, which grows to 1.5 m in length. Found from southern New South Wales to southern Western Australia, the Draughtboard Shark is a wide-ranging species. These shy sharks spend most of the day hidden in ledges or under seaweed. Like a number of catsharks they can swell their bodies by ingesting water, this is a defensive ploy that helps to lock them in a ledge so that predators can't dislodge them.

Handfish

Handfish are very weird creatures with hand-like fins, which they use to walk across the bottom. Only found in Australia, 14 species of these strange fish have so far been discovered, but all are rare, and a few are also close to extinction due to habitat loss and invasive species. Handfish are a member of the anglerfish family, as they have a head lure, but they don't seem to use it to attract prey. The only handfish seen by divers is the Spotted Handfish *(Brachionichthys hirsutus),* which grows to 12 cm in length. This species is found off Tasmania and seems to prefer muddy bottoms in bays and estuaries.

The rare and very elusive Spotted Handfish.

Frogfish

Frogfish are the cutest members of the anglerfish family. These endearing fish have great camouflage, wonderful colours, use their fins to walk and also have fishing-rod-like head lures, which they use to attract prey. Australia is home to half the known frogfish, around 24 species, and many are endemic to our southern waters.

Painted Frogfish.

A very unique Australian species is the Tasselled Frogfish (*Rhycherus filamentosus*). This species has the best camouflage of any frogfish, with its body covered in weedy filaments that closely match the algae it sits on. Found off Victoria, Tasmania and South Australia, the Tasselled Frogfish grows to 23 cm in length, but with its great camouflage this is a fish that is very difficult to find.

Striate Frogfish.

A more common species seen in tropical and subtropical waters is the Striate Frogfish (*Antennarius striatus*). This species is also known as the Hairy Frogfish, as some are covered in hair-like filaments, but most have

Tasselled Frogfish.

a distinctive striped pattern. Often found on sandy bottoms in bays, the Striate Frogfish reaches a length of 20 cm. Another species found in the same range is the Painted Frogfish (*Antennarius pictus*). This species reaches 16 cm in length and can be found in a wide range of colours from brown to orange to yellow and even red.

> Frogfish are ambush predators that usually feed on small fish. Each species has a different shaped lure attached to their head rod, some look like shrimps, others like worms. They flick these lures back and forth to attract prey, and once a fish gets close it is sucked into the frogfish's mouth in a lightning-fast attack that has been timed at only six milliseconds!

Clingfish

Hiding on seaweeds, sponges and even feather stars, clingfish are easily overlooked. Clingfish are small scaleless fish that are often decorated with bright colours. Most cling to objects, using a sucking disk formed by modified pelvic fins. In Australia 28 species of clingfish have been identified, with the family also containing a few members with eel-like bodies known as shore eels.

One of the most widespread members of this family is the Tasmanian Clingfish (*Aspasmogaster tasmaniensis*), which is found in all the southern states. This species has a pretty striped pattern, but its base colour can vary from green to pink. Best seen at night, the Tasmanian Clingfish grows to 8 cm in length. Clinging to sponges, seaweeds and kelp, the Eastern Cleaner Clingfish (*Cochleoceps orientalis*) is a species only found off New South Wales. This species grows to 5 cm in length and has been observed cleaning other fish.

Eastern Cleaner Clingfish.

Tasmanian Clingfish.

Ghostpipefish

Ghostpipefish are very delicate small fish that like to avoid predators by hiding in seagrass, corals and even amongst leaf litter. They are usually well camouflaged, and to aid their concealment they change colour to match their habitat. Ghostpipefish have an elongated snout and feed on small shrimps. Australia is home to four species of ghostpipefish, which are only found in tropical to warm-temperate waters.

The two ghostpipefish most often found in Australia are the Ornate Ghostpipefish (*Solenostomus paradoxus*) and the Robust Ghostpipefish (*Solenostomus cyanopterus*). The Ornate Ghostpipefish is the prettiest of the two, coming in a range of ornate colours and covered in spikey filaments. This species grows to 10 cm in length and is often found in pairs. The Robust Ghostpipefish is much plainer, but much larger, reaching a length of 15 cm. This species is often a drab brown colour, which aids its camouflage to make it look like a dead leaf. Both species are mostly seen in the warmer months, and where they disappear to during winter is a mystery.

The two most common ghostpipefish found in Australia are the Ornate Ghostpipefish (above) and the Robust Ghostpipefish (below).

Seahorses

Seahorses are members of the Syngnathidae family, which also includes the seadragons, pipehorses and pipefish. Australia has the world's largest variety of these delicate creatures, with around 122 species identified. All members of this family have a tube-like snout and armoured body plates, but each subfamily has other features

Australia is home to a large number of seahorses, including the Bigbelly Seahorse (above) and the Great Seahorse (right).

that make them unique. The seahorses have a prehensile tail that allows them to grip to seaweeds, corals and sponges. Found in both tropical and temperate waters, Australia is a great place to encounter seahorses.

The largest member of this family found in Australia is the Bigbelly Seahorse (*Hippocampus abdominalis*). Growing to a length of 35 cm, this large seahorse is found from South Australia to New South Wales. Distinguished by its large size and big stomach, divers generally encounter this species in sheltered bays. Another species found in Australia's temperate seas is the Shorthead Seahorse (*Hippocampus breviceps*). This small species varies greatly in colour and usually has elaborate head filaments. Only 10 cm long, the Shorthead Seahorse is sometimes hard to find, clinging to similar-coloured seaweeds.

White's Seahorse (*Hippocampus whitei*) is endemic to Australia and only found off the east coast. This species is most common off New South Wales, found in sheltered bays and estuaries. Like many seahorses, White's Seahorse varies greatly in colour,

The Shorthead Seahorse (above) is a small seahorse only found in southern Australia, while White's Seahorse (right) is only found off eastern Australia.

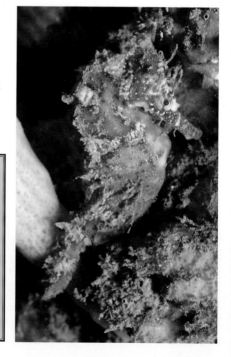

from plain grey to bright yellow, and grows to 20 cm in length. Another species occasionally found off the east coast is the Great Seahorse (*Hippocampus kelloggi*). Usually found in deep water, they sometimes venture onto shallow reefs. The Great Seahorse grows to 23 cm in length and is generally a yellow or orange colour.

Seahorses, and in fact all members of the Syngnathidae family, have a very strange reproduction strategy, with the males getting pregnant. When breeding, the female seahorse lays her eggs into a brooding pouch on the male's stomach. He then fertilises them and carries them until they hatch. Seadragon males stick their eggs on their tail, while pipefish males have a semi-enclosed brood pouch on their tail.

Seadragons

Australia is home to a very unique branch of the Syngnathidae family called seadragons. These strange creatures are an elongated version of a seahorse, covered in weedy growths. Three species are found in southern Australia, but only two are seen by divers. These well-camouflaged fish live around seagrasses and kelp, feeding on tiny mysid shrimps.

The most widespread species is the Weedy Seadragon (*Phyllopteryx taeniolatus*), that grows to 45 cm in length. Found from central New South Wales through to southern Western Australia, these spectacular fish vary greatly in colour depending on the depth and locality. Weedy Seadragons are not an easy fish to find, as even with their garish colours they still manage to blend into their kelp home. The Leafy Seadragon (*Phycodurus eques*) is even harder to find with its leaf-like growths giving it incredible camouflage. This species is only found from western Victoria to southern Western Australia and reaches a length of 35 cm.

Leafy Seadragon.

Weedy Seadragon.

Pipehorses

A cross between a seahorse and pipefish, pipehorses are rarely seen, and the less known members of the Syngnathidae family. Elongated like a pipefish, but with a prehensile tail, pipehorses are generally small and well camouflaged. A number of pipehorse species are found in tropical and temperate seas around Australia, with the Sydney Pygmy Pipehorse *(Idiotropiscis lumnitzeri)* one of the better known species.

Extremely well-camouflaged, the Sydney Pygmy Pipehorse is a difficult fish to find.

Found off a small section of the New South Wales coastline, these small fish are best seen off Sydney. Sometimes found in small groups clinging to weeds and algae, the Sydney Pygmy Pipehorse reaches a length of 6 cm.

Pipefish

Pipefish are prevalent around Australia, but being small, camouflaged and often very cryptic they are easily overlooked. These elongated fish are found on sand, reef and weed environments and mainly feed on small shrimps. Most pipefish are bottom dwellers, snaking across the bottom on their belly, but a few swim and live in caves.

Ocellate Pipefish.

The Great Barrier Reef is home to a good variety of pipefish that reside on coral heads. One widespread reef species is the Ocellate Pipefish (*Corythoichthys ocellatus*). Like a lot of reef pipefish this small fish is covered in a camouflage pattern of lines and spots. Only growing to 12 cm in length, groups of Ocellate Pipefish are often found together on a single coral head.

In southern Australia pipefish are often more difficult to find. The Ringback Pipefish (*Stipecampus cristatus*) is a temperate species found off Victoria, Tasmania and South

Ringback Pipefish.

Australia. This pipefish has a short blunt snout, and with a mottled colour pattern it tends to blend in perfectly with rubble bottoms. Growing to 25 cm in length, the Ringback Pipefish is often found under jetties. A subtropical species found off both the east and west coast is the Tiger Pipefish (*Filicampus tigris*). This large pipefish grows to 35 cm in length and lives on sandy bottoms in sheltered bays and estuaries.

The Tiger Pipefish lives in bays and estuaries.

Scorpionfish

The scorpionfish family is very diverse, with more than 100 species found in Australian waters. Lionfish, ghouls and fortesques are just a few of the subfamilies in this group that all share a common trait of venomous spines. The toxicity of this venom varies from species to species, with some lethal to humans and others

Caledonian Ghoul.

causing mild discomfort, but all should be avoided and never touched or handled. Most scorpionfish are well camouflaged, which can make them difficult to find and more likely to be encountered by accident. Scorpionfishes eat a wide variety of prey, and while most are ambush predators, others stalk the sea floor in search of a meal.

One of the most bizarre members of this family, only found in southern Australia, is the Goblinfish (*Glyptauchen panduratus*). This endemic species has a square head, a red ring around its eye and a humpback. The Goblinfish grows to 20 cm in length and is mostly active at night.

Many scorpionfish look very similar to each other. They sit on the bottom and are well camouflaged with their skin colouration and shaggy beards, so identifying individual species can be difficult. One of the easier species to identify is the Eastern Red Scorpionfish (*Scorpaena jacksoniensis*). Found off the east coast, but most common off New South Wales, the Eastern Red Scorpionfish is always red in colour and grows to 40 cm long. One of the more distinctive tropical species is the False Stonefish (*Scorpaenopsis diabolus*). This species comes in a variety of colours that make it look like a rock. The False Stonefish has a large head and humpback, and grows to 30 cm in length.

The ghouls or stingerfish are another group of scorpionfish that are found in tropical waters.

Eastern Red Scorpionfish.

False Stonefish.

Goblinfish.

Several of these unattractive fish are found in Australian waters, with the Caledonian Ghoul (*Inimicus caledonicus*) the most common. Growing to 25 cm in length, this species is found off Queensland and has a wide variety of common names, including Demon Stinger, Bearded Ghoul and Demon Stingerfish. The Caledonian Ghoul is often found buried in the sand with its lethal spines on display.

Gurnard Perch

Australia's temperate waters are home to a subfamily of scorpionfish with large heads called gurnard perch. The Common Gurnard Perch (*Neosebastes scorpaenoides*) is a widespread species found from southern New South Wales to South Australia. Found on rocky reefs, the Common Gurnard Perch grows to 40 cm and has very prominent dorsal spines. It is the most prevalent member of the gurnard perch family, but 13 other species are found in Australian waters.

The Common Gurnard Perch has a large eye and large spines.

Stonefish

The most dangerous fish in Australian waters are the stonefish. These well-camouflaged fish have a row of 13 lethal spines in their dorsal fin that are fed by twin venom sacks. Australia is home to four of these fish, with the Reef Stonefish (*Synanceia verrucosa*) the species most often seen by divers. Hiding among coral and rubble, the Reef Stonefish is generally well concealed as it awaits prey. The Reef Stonefish is found in tropical waters and grows to 38 cm in length.

The Reef Stonefish is one of the most dangerous fish in Australia.

Velvetfish

Closely related to the scorpionfish, velvetfish have sandpaper-like skin and many species have venomous spines. These bottom-dwelling fish are generally well camouflaged, hiding among rocks, weeds and debris. Australia is home to a great variety of these unusual fish, with around 21 species so far identified.

A number of velvetfish are found in Australia's southern waters, with the Southern Velvetfish (*Aploactisoma milesii*) the most common. This species grows to 23 cm in length and varies greatly in colour. To aid in camouflage Southern Velvetfish can also become covered in algae. A number of new and endemic velvetfish have been found in Australian waters during the last decade, including the Goatee Velvetfish (*Pseudopataecus carnatobarbatus*). This pretty velvetfish is only found off

Goatee Velvetfish.

Southern Velvetfish.

the north-west coast of Western Australia. Growing to a length of 10 cm, the Goatee Velvetfish varies greatly in colour, but always has beard-like growths on its chin.

Prowfish

Only found in southern Australian waters, the prowfish are very weird creatures that look like sponges and shed their skin as they grow. Looking like a sponge allows prowfish to blend into the bottom and ambush shrimps and crabs. Three species of prowfish are found in Australian waters, but only one is regularly seen.

One of the strangest fish found in Australia is the Red Indian Fish.

The Red Indian Fish (*Pataecus fronto*) is the only member of this family regularly encountered by divers. This strange fish obtained its unusual name as its high dorsal fin looks very similar to a North American Indian chief's headdress. While found off southern Western Australia and South Australia, it is most prevalent in sponge gardens off New South Wales. Growing to a length of 35 cm the Red Indian Fish is quite large, but well camouflaged they are not easy to find.

Weedfish and Snake Blennies

Only found in temperate waters, weedfish and snake blennies look quite different but are grouped together in the one family. Both like to hide among seaweeds, rocks and other bottom debris, and have elongated bodies to allow them to easily slide in and out of holes. The weedfish typically have larger fins than snake blennies, most being

Always found hiding in seaweed is the Yellow-crested Weedfish.

weed-like. Both groups also have camouflaged skin patterns to aid with concealment. Australia is home to a good variety of these fish, with around 32 species recognised.

One of the prettiest members of this family is the Yellow-crested Weedfish (*Cristiceps aurantiacus*). This species is found in areas of seaweed and kelp around southern Australia and varies in colour from yellow to orange. The Yellow-crested Weedfish has a flag-like dorsal fin and grows to 18 cm in length. Much less often seen

is the Dusky Snake Blenny
(*Ophiclinus antarcticus*). This
eel-like fish grows to 17 cm in
length and varies in colour from
yellow to brown. Found off
South Australia and southern
Western Australia, the Adelaide
Snake Blenny is not an easy
species to find as it likes to hide
under rocks and debris.

Dusky Snake Blenny.

Dragonets

Dragonets lack scales and instead have a tough skin covered in mucus that gives off
a strong odour, a reason they are also known as stinkfish. Found around Australia,
dragonets are bottom dwellers that walk with their fins while looking for small
invertebrates to eat. Most dragonets have camouflaged skin patterns, but a few are
very brightly coloured and rely on their foul odour to keep them safe from predators.
Around Australia 48 species of dragonets have been identified, including many
endemic species.

The Finger Dragonet walks with the aid of modified fins.

Morrison's Dragonet (above) and Painted Stinkfish (below).

A number of dragonets are found in Australia's temperate waters, including the attractive Painted Stinkfish (*Eocallionymus papilio*). This species has a mottled skin pattern with the males more colourful than the females. The Painted Stinkfish grows to 13 cm in length and is generally seen on sandy bottoms looking for food. Also found on sandy bottoms in tropical waters is the Finger Dragonet (*Dactylopus dactylopus*). Reaching a length of 30 cm, it is one of the largest dragonets. The Finger Dragonet walks with the aid of modified pelvic fins that look like fingers. It also uses these fingers to dig for food. A number of small dragonets also live among corals, including Morrison's Dragonet (*Neosynchiropus morrisoni*). This colourful fish grows to 8 cm in length and is found in Australia's tropical seas.

Blennies, like the Filamentous Blenny *(Cirripectes filamentosus)* live in holes in the bottom.

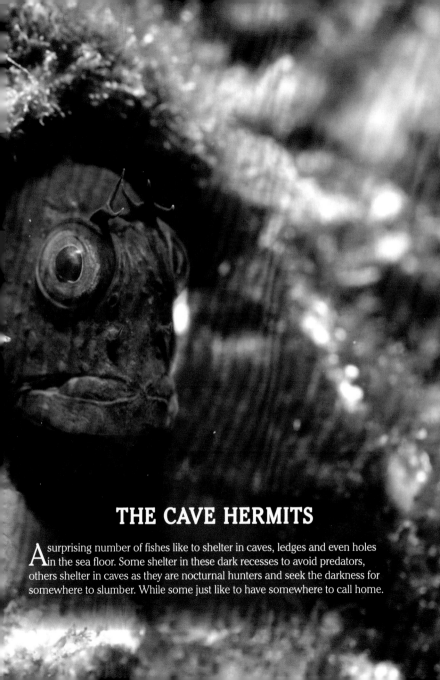

THE CAVE HERMITS

A surprising number of fishes like to shelter in caves, ledges and even holes in the sea floor. Some shelter in these dark recesses to avoid predators, others shelter in caves as they are nocturnal hunters and seek the darkness for somewhere to slumber. While some just like to have somewhere to call home.

Blind Sharks

Many small shark species are eaten by their larger relatives, so they are quite shy and like to hide in caves and ledges by day, only nervously emerging at night to feed. One family of shy cave dwellers are the blind sharks.

The blind shark family is very small, only containing two members that are endemic to Australia. These small sharks are not blind but received their strange name as they close their eyes when removed from the water. The most common member of this family is simply called the Blind Shark (*Brachaelurus waddi*). This species is only found in New South Wales and southern Queensland and reaches a length of 1.2 m. Found on rocky reefs, by day a tail hanging out of a hole is often all that is seen of a Blind Shark. The only other member of this family is the much rarer Colclough's Shark (*Brachaelurus colcloughi*). A grey colour, this species is only found off southern Queensland and northern New South Wales and is easily mistaken for the more common Brown-banded Bamboo Shark.

Blind Shark.

Colclough's Shark.

Epaulette and Bamboo Sharks

Greater Bamboo Shark.

Another shy family of sharks are the epaulette and bamboo sharks. Three members of this family are found in Australia's tropical waters. These sharks have slender elongated bodies, which allow them to crawl into holes and crevasses. While placed in the same family, bamboo sharks are generally larger and have plainer colours, with the Greater Bamboo Shark *(Chiloscyllium magnus)* a good example of this. This species only has brown bands when young, with the adults typically a grey colour. Epaulette sharks on the other hand have elaborate body patterns. The only member of this family regularly seen in tropical Australia is the Epaulette Shark *(Hemiscyllium ocellatum)*. Like all epaulette sharks this species has a large spot above the pectoral fins that looks like an epaulette. With a preference for shallow water it is common on reef flats in only 1 m of water.

Epaulette Shark.

Epaulette sharks are also known as walking sharks, as they walk across the bottom on their fins when looking for food. They can also walk on land, as living in shallow water they sometimes get trapped in rock pools at low tide and will walk from pool to pool in the hunt for prey.

Giant Moray.

Moray Eels

The morays are one of the largest eel families, with 62 species found in Australian waters. Potentially dangerous, because of their razor-sharp teeth, most morays are shy and reclusive, preferring to hide in their hole than attack a diver or snorkeller. The only time when they are really dangerous is when they have been hooked by an angler and fight to break free. Most moray eels are nocturnal, spending the day resting in a cave or ledge and emerging at night to hunt prey. Most feed on fish, but many will also take invertebrates. Moray eels are mostly found in tropical waters, but a few species range into temperate seas.

One of the largest morays found in Australia is the Giant Moray (*Gymnothorax javanicus*). This tropical species is prevalent on the Great Barrier Reef and reaches a length of 2.5 m. Typically a drab brown colour, the Giant Moray is easily identified by its large size. Another common tropical moray is the Honeycomb Moray (*Gymnothorax favagineus*). This species has a pretty black-and-white pattern that can vary quite dramatically, sometimes dominated by white and other times by black. This species grows to 2 m in length and is also found in subtropical regions.

Australia's most spectacular moray, which is only found in the warm-temperate seas off New South Wales, is the Mosaic Moray (*Enchelycore ramosa*). This species has a mouth full of needle-like teeth and a brilliant mosaic pattern across its body. The Green Moray (*Gymnothorax prasinus*) is the only moray regularly seen in Australia's

A large variety of moray eel species are found in Australia, including the Green Moray (left) and Honeycomb Moray (right).

The Mosaic Moray is a spectacular eel.

temperate seas. Found off New South Wales and southern Western Australia, this species grows to 1 m in length. The Green Moray is not always green, as this eel can vary in colour from yellow to brown.

Moray eels generally eat their preferred diet of fish whole, and not having hands makes this a tricky meal to swallow. To help get their food down they have a second hidden set of jaws located in the throat, which has been likened to the creature in the *Alien* movies. When swallowing a fish this second set of teeth reach forward to drag the food into the eel's stomach.

Eastern Toadfish.

Toadfish

Toadfish have had their name hijacked by a number of fish species that look nothing like a toad, but with a wide mouth and flattened head toadfish really do look like a toad. This strange family of fish live in holes and ambush prey that encroach close to their lair. While they may look unattractive, toadfish are the best parents of any fish species, with the male guarding his clutch of eggs until they hatch and then looking after the young until they are big enough to fend for themselves.

Australia is home to nine species of toadfish, with the Eastern Toadfish (*Batrachomoeus dubius*) the most common member of the family. This camouflaged fish grows to 30 cm in length and is found off the east coast of Australia but is most abundant off New South Wales. A rarer species sometimes seen in the tropical waters of Australia is the Banded Toadfish *(Halophryne diemensis).*

This species reaches a length of 20 cm and is best seen on Ningaloo Reef. All toadfish have sharp spines and a number are venomous, so they should never be touched or handled.

Banded Toadfish.

Bizarre-looking Common Pineapplefish are often found in small groups in caves.

Pineapplefish

One of the most unusual fish families found in Australian waters is the pineapplefish. The name kind of gives it away as these bizarre fish look like a pineapple. A base yellow colour trimmed with black diamond patterns, pineapplefish have a hard exoskeleton and short sharp spines for defence. Two species of pineapplefish are found in Australia, with the Common Pineapplefish *(Cleidopus gloriamaris)* the only prevalent species.

Growing to a length of 25 cm, Common Pineapplefish are found in warm-temperate seas off the east and west coast of Australia. By day they shelter in caves and ledges, often in small groups, and have a strong site fidelity, using the same hiding spot for years. Divers can get very close to study these strange fish and if you listen closely you may even hear them croak!

> Pineapplefish not only look strange, but also have headlights. Located below the eye these fish have a bioluminescent organ that is covered by a lid. At night they open this lid and let the organ glow its eerie green light. This glow attracts small shrimps, which the pineapplefish then feasts on.

Soldierfish and Squirrelfish

Resting in caves by day and hunting the reef at night, soldierfish and squirrelfish are red-coloured fish classed together in the same family, Holocentridae. These fish look very similar, however, there are subtle differences between the two, with soldierfish having larger eyes and a blunt head. Both have spines on their gill cover, which are venomous in some species. Australia is home to 32 species of soldierfish and squirrelfish, and all are found in tropical waters.

Crimson Soldierfish school in caves.

The largest member of this family is the Giant Squirrelfish (*Sargocentron spiniferum*). These fish grow to 45 cm in length and with a very large gill spine they are also known as Sabre Squirrelfish. Giant Squirrelfish are usually found singularly, or in small groups of two or three, unlike most other species in this family that like to school. One species always found in schools is the Crimson Soldierfish (*Myripristis murdjan*). This species is prevalent on inshore reefs from northern New South Wales to central Western Australia and grows to 23 cm in length.

The Giant Squirrelfish is the largest member of this family.

Gropers and Soapfish

Most people think of gropers as very large fish, and while there are some huge gropers, this vast family also contains many subfamilies of small fishes. Gropers are just one member of the Serranidae family, which also contains the rockcods, basslets, seaperch and soapfish. Shared characteristics of this family include small scales, an indistinct lateral line and a rounded caudal fin. In Australian waters there are

Goldspotted Groper.

148 members of this family, with the gropers and soapfish often found dwelling in caves.

Most gropers are only found in tropical waters, but Australia's largest groper

Malabar Groper.

Potato Cod.

species, the Queensland Groper (*Epinephelus lanceolatus*), also resides in warm-temperate zones. These giant fish grow to 2.7 m in length and like to hide in caves and shipwrecks. Like most gropers they are a shy animal that rarely let a diver get close. The Potato Cod (*Epinephelus tukula*) is also shy, but a famous group of these fish have been handfed for years at a site called Cod Hole on the Great Barrier Reef and are now very tame. A tropical species, the Potato Cod, grows to 2 m in length.

A very widespread species is the Goldspotted Groper (*Epinephelus coioides*). This groper reaches a length of 1 m and has distinctive brown or gold spots. Found from central New South Wales to central Western Australia, the Goldspotted Groper is not nearly as shy as other members of this family and some will allow divers to get very close. Another large member of this family found in Australia's tropical waters is the Malabar Groper (*Epinephelus malabaricus*).

Queensland Groper.

This species has a pretty spotted pattern and grows to 2.3 m in length.

Soapfish are another tropical member of this family that like to shelter in caves, with the most common being the Barred Soapfish (*Diploprion bifasciatum*). This species ranges into the warm-temperate zone and grows to 25 cm long. The Lined Soapfish (*Grammistes sexlineatus*) is a tropical species, but rarely seen. This fish grows to 27 cm in length and its stripes can vary in colour from white to yellow.

The two most common soapfish are the Barred Soapfish (left) and the Lined Soapfish (right).

Longfins

The longfin family is due for a review as it contains several subfamilies that are quite different from each other. Included in this family are the small scissortails and hulafish, and the larger devilfish. While all these fish do have long fins, they also have different body shapes and feed on different prey. However, they all like to hangout in caves.

Eastern Blue Devilfish.

In Australian waters are 23 species of longfins, with the best known members of this family being the devilfish. While the other members of this family are observed darting about, devilfish are more stationary and like to sit on the bottom, propped up by their oversized fins. The Southern Blue Devilfish (*Paraplesiops meleagris*) is a prevalent

The wonderful Southern Blue Devilfish.

species from Victoria to central Western Australia. With its wonderful blue polka-dot pattern it is a very striking fish and divers can often get very close for photos. The Southern Blue Devilfish grows to a length of 33 cm and dwells in caves on rocky reefs. The Eastern Blue Devilfish (*Paraplesiops meleagris*) is even more colourful, with its white bands and yellow fins. Only found off New South Wales, the Eastern Blue Devilfish grows to 40 cm in length.

Cardinalfish

The cardinalfish are a family that like to shelter in caves and among corals. These small fish are usually found in schools, but some species seem to prefer their own company. Cardinalfish feed on invertebrates, plankton and even small fish. Australia is home to a diverse range of cardinalfish, around 125 species, and most are found in tropical waters.

Many cardinalfish have striped patterns, including the Fiveline Cardinalfish (*Cheilodipterus quinquelineatus*). This tropical species is abundant on the Great Barrier Reef, hiding among coral branches. The Fiveline Cardinalfish grows to 12 cm in length and is generally found in small groups. The Sydney

The Fiveline Cardinalfish is a common reef species.

Sydney Cardinalfish (left) and a Plain Cardinalfish (right) with a mouthful of eggs.

Cardinalfish (*Ostorhinchus limenus*) is another striped member of this family that is endemic to the east coast of Australia. Often found in caves off Sydney, this species grows to 14 cm in length. The Plain Cardinalfish (*Ostorhinchus apogonides*) is similar to a number of species with blue eye stripes. It is found throughout Queensland and reaches a length of 10 cm.

> Cardinalfish have one of the stranger forms of reproduction, a strategy called mouth brooding. After the female lays her eggs and the male fertilises them, the male, and in a few cases the female, takes the eggs into its mouth and keeps them there until they hatch. The parent fish doesn't eat until the young hatch, which can take up to 30 days.

Bullseyes

Sheltering in caves by day in large schools, bullseyes venture across the reef at night to feed on crustaceans and cephalopods. This family of fish have large eyes to aid their night-time hunting activities. Australia is home to 13 bullseyes, with species found in both

Black-tipped Bullseye.

tropical and temperate waters. The Black-tipped Bullseye *(Pempheris affinis)* is one of the more common members of this family. Found throughout New South Wales and southern Queensland, the Black-tipped Bullseye grows to 15 cm in length and is easily identified by the black tips on its fins.

Blennies

With big eyes and thick lips, blennies are very humorous and attractive little fish. Blennies are bottom-dwelling fish that live in holes, only leaving their homes to feed, fight and mate. These elongated scaleless fish feed on algae, zooplankton and small invertebrates. Blennies are more typical in tropical waters and around 92 species are found in Australia.

Australian Combtooth Blenny.

One of the prettiest members of this family is the Leopard Blenny *(Exallias brevis)*. Growing to 14 cm in length it is bigger than most blennies. This tropical species likes to hide in branching hard corals, darting from lookout to lookout in a quest for food. Many blennies leave the safety of their home when looking for food and prop themselves on a perch. The Australian Combtooth Blenny *(Ecsenius australianus)* is a species that is often seen sitting out in the open. Only growing to 6 cm in length, this species is widespread on the Great Barrier Reef.

Brown Sabretooth Blenny.

The Leopard Blenny is one of the prettiest members of this family.

The Triplespot Blenny (*Crossosalarias macrospilus*) is another tropical species that reaches a length of 10 cm. These pretty fish are good parents, and like most blennies the male guards the eggs in his hole until they hatch. One of the more abundant Australian species is the Brown Sabretooth Blenny (*Petroscirtes lupus*). The sabretooth blennies have long sharp fangs in their lower jaw and will bite if handled. This species grows to 12 cm in length and is found from central Western Australia to southern New South Wales.

Triplespot Blenny.

The Coral Rockcod *(Cephalopholis miniata)* is a very colourful reef fish.

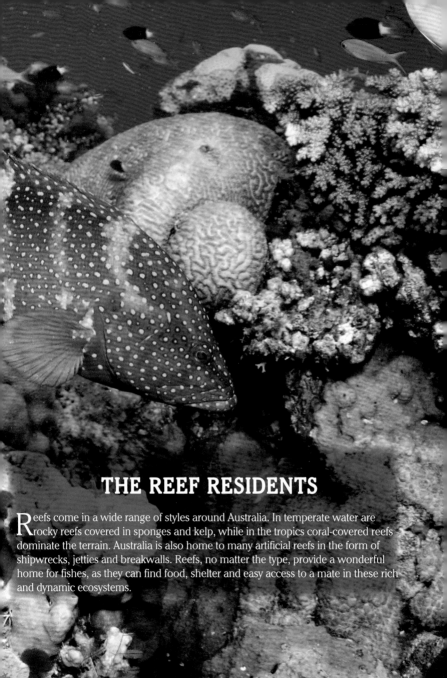

THE REEF RESIDENTS

Reefs come in a wide range of styles around Australia. In temperate water are rocky reefs covered in sponges and kelp, while in the tropics coral-covered reefs dominate the terrain. Australia is also home to many artificial reefs in the form of shipwrecks, jetties and breakwalls. Reefs, no matter the type, provide a wonderful home for fishes, as they can find food, shelter and easy access to a mate in these rich and dynamic ecosystems.

Horn sharks

Many shark species visit reefs to feed, but a handful spend almost all their time in this environment. The horn sharks are a very docile family of sharks, and also a very old one, still retaining spines on their dorsal fins which were common in their ancestors. These reef dwellers have very small teeth, arranged in a plate-like manner,

Horn sharks are quite primitive sharks with dorsal spines. The only common species in Australia are the Crested Horn Shark (above) and Port Jackson Shark (below).

which they use to crush the shells of prey. Horn sharks mostly feed on molluscs and crustaceans, but will also feed on fish. Australia is home to three species of horn sharks, with the Port Jackson Shark *(Heterodontus portusjacksoni)* the most common member of the family.

The Port Jackson Shark is the largest of all the horn sharks, growing to 1.6 m in length. Found throughout Australia's temperate waters, they are most prevalent off southern New South Wales. These sharks are easily approached by divers, and during the winter months large numbers can be seen when they gather to breed. New South Wales is also home to another species from this family, the Crested Horn

Shark (*Heterodontus galeatus*). This species looks very similar to its cousin, but has a banded pattern and is smaller, only growing to 1.3 m in length. It is much rarer and likes to shelter in caves.

Nurse Sharks

Another shy reef shark that likes to rest in caves and gutters is the Tawny Nurse Shark *(Nebrius ferrugineus)*. A member of the small nurse shark family, it is the only representative in Australian waters. Nurse sharks are nocturnal hunters, preying on a range of fish, cephalopods and crustaceans. The Tawny Nurse Shark grows to 3.2 m in length and is found throughout Australia's tropical seas. A docile species they have small teeth, but divers have been rammed by a startled Tawny Nurse Shark that has been accidently woken.

Tawny Nurse Shark.

Sergeant Bakers

The sergeant bakers are elongated reef fish that like to rest on the bottom when not ambushing prey. Most species live in deep water, and five species have been recorded around Australia. The most common and widely distributed member of this family is the Sergeant Baker *(Hime purpurissatus)*. These

Sergeant Baker.

large reddish-coloured fish grow to 60 cm in length and are found on rocky reefs. Endemic to southern Australia, Sergeant Bakers are found from southern Queensland to central Western Australia.

Lizardfish

Lizardfish are generally observed sitting motionless on reefs and looking very dull. However, these elongated fish are ferocious hunters, launching off the bottom to grab small fish with their needle-like teeth in lightning fast attacks. Mostly found in tropical waters, Australia is home

Twospot Lizardfish.

to 27 species of lizardfish. The Twospot Lizardfish *(Synodus binotatus)* is a tropical species and identified by the two spots on its snout. This species grows to 18 cm in length and pairs are often found sitting together.

Pacific Trumpetfish.

Trumpetfish

This small family of fish contains only two members, and one is very common in Australia's tropical and subtropical waters. The Pacific Trumpetfish *(Aulostomus chinensis)* is an elongated fish that grows to 80 cm in length. This species can change its colour, but is generally brown or yellow. Pacific

Trumpetfish are very sneaky animals that use larger fish, and sometimes divers, to get close to small fish so they can catch and eat them.

Lionfish

Members of the vast and varied Scorpaenidae family, lionfish are common on Australia's tropical and subtropical reefs. While very attractive with their feather-like fins, all lionfish are venomous and

Common Lionfish.

very efficient predators. Feeding on small fish, they use their fins to round up prey. The largest and most abundant member of this family is the Common Lionfish *(Pterois volitans)*. This species grows to 35 cm in length, and safe in the knowledge that few predators will tangle with its venomous spines, the Common Lionfish boldly stalks the reef by day.

Twinspot Lionfish.

Most other lionfish are a little more cryptic, like the Zebra Lionfish *(Dendrochirus zebra)*. This species grows to 18 cm in length, but is usually much smaller. Zebra Lionfish are found on coral reefs and rocky reefs from central Western Australia to central New South Wales. One of the rarer members of this family found in tropical waters is the Twinspot Lionfish *(Dendrochirus biocellatus)*. This species has very distinctive eye spots on its second dorsal fin and appears to be nocturnal.

Zebra Lionfish.

Rockcods, Seaperch and Basslets

These fishes are all members of the complex Serranidae family. The gropers and soapfish are the shyer cave hermit members of this family, while the rockcods, seaperch and basslets are reef dwellers. The rockcods are the most diverse group.

Harlequin Fish.

One of the best known members of the rockcod family is the Common Coral Trout *(Plectropomus leopardus)*. Australia is home to a number of coral trout species, and all are popular with anglers. This tropical species grows to 75 cm in length. Another frequently found tropical rockcod is the Barramundi Rockcod *(Chromileptes altivelis)*. This fish has a small pointed head and grows to 70 cm in length.

Most rockcods are quite small, such as the Honeycomb Rockcod (*Epinephelus merra*). This tropical species grows to 28 cm in length, and like many small rockcods seems to spend most of the day just sitting on the bottom. While most rockcods are tropical, the endemic Harlequin Fish (*Othos dentex*) prefers temperate waters off southern Western Australia and South Australia. These pretty reef fish reach a length of 76 cm.

These species show the variety in this family, the Redbar Basslet (above), Barramundi Rockcod (below), Banded Seaperch (bottom left) and Butterfly Perch (bottom right).

Seaperch are the cool-water version of a rockcod, and a number of species are found in Australia's southern waters. The Banded Seaperch (*Hypoplectrodes nigroruber*) is one of the more common members of this subfamily and found from central New South Wales to southern Western Australia. Mostly seen sitting on the bottom, this species grows to 30 cm in length. The perch are another

closely related subfamily, which also prefer temperate waters. The Butterfly Perch (*Caesioperca lepidoptera*) is the best known member of this family and is often found in dense schools. Found from southern Western Australia to southern New South Wales, the Butterfly Perch grows to 40 cm in length.

The basslets are the prettiest member of the Serranidae family. These colourful fish are found in large numbers on coral

reefs, with a very colourful male looking after his harem of slightly plainer females. A variety of basslet species are found in Australia's tropical waters. The Redbar Basslet (*Pseudanthias rubrizonatus*) is prevalent throughout Queensland and grows to 12 cm in length. In this species the male has a striking red panel on its side, while the females are reddish-pink. The Orange Basslet (*Pseudanthias squamipinnis*) is regularly seen on the Great Barrier Reef. The females of this species are orange, while the males are a lovely purplish-pink colour.

Bigeyes

With very large eyes it is easy to see where the bigeyes got their name. These red-coloured reef dwellers are nocturnal feeders, and while some linger in caves, others simply swim around the reef. Often found in schools or small groups, nine species

Lunartail Bigeye.

of bigeyes are found in Australia's tropical seas. The Lunartail Bigeye (*Priacanthus hamrur*) is the most abundant member of this family. This species grows to 40 cm in length and is found on shallow reefs to depths of 250 m.

Threadfin Breams

Although this family is named after the threadfin breams, this complex group also includes the whiptails and spinecheeks. The members of this family are small reef dwellers, and many are only found in deep, tropical waters. Australia is home to 36 species of threadfin breams, the most common of which is the Two-line Monocle Bream (*Scolopsis bilineata*). This species grows to 23 cm long and is a type of spinecheek, possessing a sharp spine below the eye. The Paradise Threadfin Bream (*Pentapodus paradiseus*) is a colourful fish that grows to 30 cm long.

Paradise Threadfin Bream.

Two-line Monocle Bream.

Emperors

The emperors are reef fish that feed on a variety of small invertebrate species. Found in tropical waters, Australia is home to 28 species of these medium-sized reef fish. One of the prettiest members of this family is the Redthroat Emperor *(Lethrinus miniatus)*. Regularly seen on coral reefs from central Western Australia to northern New South Wales, the Redthroat Emperor grows to 90 cm in length and is popular with anglers.

The Redthroat Emperor is a popular fish with anglers.

Butterflyfish

Some of the most beautiful fish seen on Australia's coral reefs are the multi-coloured butterflyfish. Small in size, but hard to miss because of their dazzling colour range, butterflyfish are almost exclusively found in tropical waters. Butterflyfish eat a variety of food, depending on the species. Most consume small invertebrates, but others eat

A pair of Beaked Butterflyfish.

A sample of butterflyfish include the Blue-dash Butterflyfish (top), Reef Bannerfish (above left) and Square-back Butterflyfish (above right).

zooplankton, coral polyps and even sea anemones. Territorial, many butterflyfish are always found in pairs, but the ones feeding on zooplankton form into schools. Australia is home to 54 species of butterflyfish.

The Blue-dash Butterflyfish (*Chaetodon plebeius*) is a widespread species found in tropical and subtropical waters. This species grows to 12 cm in length and is always seen in pairs. Most butterflyfish have short snouts, but a few, like the Beaked Butterflyfish (*Chelmon rostratus*), have elongated snouts for feeding. These handsome fish are found in tropical and subtropical zones around Australia and reach 20 cm in length. The Reef Bannerfish (*Heniochus acuminatus*) is another widespread species that has an extended, banner-like, dorsal fin. Always found in pairs, the similar-looking Schooling Bannerfish (*Heniochus diphreutes*) prefers to school. The Square-back Butterflyfish (*Chelmonops curiosus*) is one of the few temperate-water members of this family. Found off South Australia and southern Western Australia, this species grows to 20 cm in length.

Butterflyfish, like many fish species, have a pelagic stage when young. This disperses the young over a wide range, as they are carried by ocean currents. In Australia these currents flow south, meaning juvenile butterflyfish often end up in temperate waters during summer. Unfortunately, once the ocean temperature cools in winter these northern invaders die.

Angelfish

The angelfish are very regal reef residents and are closely related to the butterflyfish, but have a spine on their operculum (gill cover). They are always brightly coloured, with some having different colours and patterns for the males, females and juveniles. Angelfish feed on a variety of food, some eating corals, others algae, invertebrates and zooplankton. Australia is home to 29 species of angelfish, and all are found in tropical waters.

Angelfish are some of the prettiest reef fish with the family including the Bicolor Angelfish (above), Regal Angelfish (below left) and Six-banded Angelfish (below right).

One of the most beautiful angelfish found on coral reefs is the Regal Angelfish (*Pygoplites diacanthus*). These multicoloured fish grow to 25 cm in length and feed on sponges and sea squirts. The Six-banded Angelfish (*Pomacanthus sexstriatus*) is one of the larger members of the family, reaching 46 cm in length. This species is always found in pairs. Many angelfish are small and shy, darting among corals. One common small species is the Bicolor Angelfish (*Centropyge bicolor*), which grows to 15 cm in length and is found in pairs or small groups.

Boarfish

Only found in temperate waters, the boarfish family is well represented in Australia with 10 species known. Boarfish have large fins and a long snout, and most species are only found in deep water. The most common and widespread species in this family is the

Longsnout Boarfish.

Longsnout Boarfish *(Pentaceropsis recurvirostris).* These striking black-and-white fish are found from central New South Wales to southern Western Australia and grow to 50 cm in length.

Morwongs

Although their body shape varies between species, the morwongs typically have thick lips and long pectoral fins. Often found sitting on the bottom, morwongs feed on small invertebrate species. Australia is home to a dozen species of morwong, and most are found in temperate waters.

The Magpie Perch *(Cheilodactylus nigripes)* is one of the morwongs seen on southern rocky reefs. This species grows to 40 cm in length, and is often observed resting on its fins. The Red Morwong *(Cheilodactylus fuscus)* is the most abundant member of this family found in New South Wales. Growing to 65 cm, it is a large fish that often gathers in schools.

Magpie Perch.

Red Morwong.

Bastard Trumpeter.

Trumpeters

The trumpeters are closely related to the morwongs and only found in temperate waters. Generally found in schools, and popular with anglers, Australia is home to four members of the trumpeter family. The most abundant member of this family is the Bastard Trumpeter *(Latridopsis forsteri).* Found from New South Wales to South Australia, these cool-water fish grow to 65 cm in length.

Seacarps

This small family of fish are only found in the Southern Hemisphere, with four species found in southern Australia. Seacarp are elongated fish that live on rocky and weedy bottoms. They have camouflaged skin patterns and often sit on the bottom. Seacarp feed on seaweeds and algae. The Western Seacarp *(Aplodactylus westralis)* is found on weed-covered reefs in Western Australia. These fish have a pretty banded and spotted pattern and grow to 35 cm long.

Western Seacarp.

Kelpfish

Kelpfish spend most of their time resting on or under seaweeds. Only found in temperate waters, Australia is home to six species of these fish that feed on small invertebrates. The only member of this family that is regularly seen by divers is the Eastern Kelpfish *(Chironemus marmoratus)*. This species is found from southern Queensland to eastern Victoria and grows to 40 cm in length.

Eastern Kelpfish.

Hawkfish

Always seen sitting on a perch, hawkfish are constantly on the lookout for predators and prey. Australia is home to 12 species of these colourful small fish that inhabit tropical reefs and feed on a range of small invertebrates and fish. The Falcon Hawkfish *(Cirrhitichthys falco)* is common on the Great Barrier Reef. Growing to 7 cm in length, this

Falcon Hawkfish.

species is often found in pairs. The Ringeye Hawkfish *(Paracirrhites arcatus)* is a widespread tropical species that reaches 14 cm in length. Generally found sitting on gorgonians and black coral trees, the Longnose Hawkfish *(Oxycirrhites typus)* is one of the prettiest members of this family. This species has

Longnose Hawkfish.

a distinctive elongated snout and grows to 13 cm long. One of the rarer Australian hawkfish is the Splendid Hawkfish *(Notocirrhitus splendens)*, which is only found off New South Wales and Queensland, and grows to 20 cm in length.

The Ringeye Hawkfish (left) is a common reef fish, while the Splendid Hawkfish (right) is quite rare.

Damselfish

The damselfish are a very large family of small reef fish. Australia is home to a diverse range of these fishes, with 140 species identified. There are three subclasses in this family, the damsels, the

Blue-green Puller congregate on corals.

Some common damselfish include the Golden Damsel (above), Headband Humbug (middle) and Sergeant Major (below).

humbugs and pullers and the anemonefish. All three are very similar in size and shape, mainly varying in their behaviour and feeding habits.

The damsels are found in both tropical and temperate waters, and are the largest members of the family, and often very territorial. The Golden Damsel (*Amblyglyphidodon aureus*) is a pretty reef resident that grows to 12 cm long. One of the more common and widespread damsels is the Sergeant Major (*Abudefduf vaigiensis*). These fish grow to 15 cm in length and are found in tropical and subtropical waters. Sergeant Major are often seen in groups on the reef, especially when breeding and defending their eggs.

The humbugs and pullers are generally smaller than the damsels, and many hide in corals. Most of these small fish feed on plankton. The Blue-green Puller (*Chromis viridis*) and Headband Humbug (*Dascyllus reticulatus*) are both common members of this family. Reaching a length of 8 cm, both live in schools that shelter in branching hard corals.

Some damselfish are very territorial, guarding a home patch from intruders, and can be very aggressive when looking after their eggs. These small fish launch kamikaze-style attacks on much larger fish, and also divers that get close to their eggs. Many a diver has left the water bitten and bleeding after an attack from a 10 cm long damselfish!

Anemonefish

Eastern Clown Anemonefish.

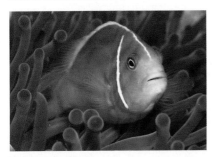

Pink Anemonefish.

White-band Anemonefish.

The anemonefish are the most popular and best known of the damselfish family. Living in a symbiotic relationship with sea anemones, these small fish have a mucus coating that stops the sea anemone stinging. While the anemonefish gets a safe home, they also protect the sea anemone from potential predators like turtles, haranguing them so they don't feast on the tentacles. Most anemonefish are found in tropical waters, but some also reside in subtropical zones. Australia is home to a dozen species of these colourful fish.

The best-known member of this family is the Eastern Clown Anemonefish (*Amphiprion percula*), the star of the film *Finding Nemo*. These pretty fish grow to 8 cm in length and are best seen on the Great Barrier Reef. The Pink Anemonefish (*Amphiprion perideraion*) is a more widespread species, found in tropical waters across Australia. A very unique Australian member of this family is the White-band Anemonefish (*Amphiprion latezonatus*). This species is endemic to southern Queensland and northern New South Wales and grows to 15 cm in length.

Anemonefish have very complex social lives, with the sea anemone dominated by a large female. She has a male partner, who looks after the eggs, but they also share their sea anemone with a group of smaller males. If the female dies, her male partner changes sex and becomes the boss, while one of the smaller males steps up to take his place.

Wrasses

The wrasse family is very complex and diverse, with some members of this family only a few centimetres long and others more than two metres in length. Within this family are many subfamilies, including the pigfish, Maori wrasse, tuskfish, cleaner wrasse and razorfish. The wrasses are found on reefs in both tropical and temperate waters, and identifying species can be tricky as males and females of the same species have

Bird Wrasse.

different colours. The males are generally larger and rule a harem of females, and if the male dies the largest female changes sex to take his place. Wrasses eat a variety of food depending on the species, including invertebrates, small fish and

The Blackfin Pigfish is a pretty wrasse species found on the Great Barrier Reef.

zooplankton. Australian waters contain a rich variety of wrasse species, with 230 identified.

Eastern Blue Groper.

Most wrasses are small and colourful, like the Leopard Wrasse (*Macropharyngodon meleagris*) and Bird Wrasse (*Gomphosus varius*). Both species are found on tropical coral reefs and reflect the diversity of this family. The Leopard Wrasse grows to 13 cm in length, while a fully grown Bird Wrasse can be 30 cm long. In southern waters the wrasses are just as colourful, like the pretty Redband Wrasse (*Pseudolabrus biserialis*). This species is only found

A variety of wrasses, Giant Maori Wrasse (above left), Harlequin Tuskfish (above right), Leopard Wrasse (below) and Redband Wrasse (bottom).

off southern Western Australia and grows to 25 cm long.

Maori wrasse inhabit coral reefs, and members of this subfamily typically have thick lips and patterns on their face like Maori tattoos. Most members of this subfamily are small, like the Redbreast Maori Wrasse (*Cheilinus*

fasciatus) that reaches a length of 36 cm. While the Giant Maori Wrasse (*Cheilinus undulatus*) is a real giant of the family, growing to 2.3 m long.

Tuskfish, with their protruding teeth, are an easily identified subfamily. Found on coral reefs, the best known member of this family is the Harlequin Tuskfish (*Choerodon fasciatus*). These multicoloured fish grow to 30 cm in length and use

their protruding dentures to lift up rocks in search of invertebrates. The pigfish may have an unappealing name, but they are an attractive subfamily in this group. The Blackfin Pigfish (*Bodianus loxozonus*) is typical of this group with a pointed snout and pretty body pattern. This species is found on the Great Barrier Reef and grows to 40 cm long.

The wrasse with the most personality would have to be the Eastern Blue Groper (*Achoerodus viridis*). Found from southern Queensland to eastern Victoria, they are most common off New South Wales. These cheeky and inquisitive fish grow to 1.2 m in length and follow divers around hoping for a meal. They love to eat sea urchins, and have very thick lips to deal with these spiny creatures.

Thornfish

Redbreast Maori Wrasse.

Only found in the cool waters of the Southern Hemisphere, thornfish are bottom-dwelling fish with a sharp spine on the operculum (gill cover).

Three species are found in Australia, but only the Thornfish (*Bovichtus angustifrons*) is common. Found off South Australia, Tasmania and Victoria, the Thornfish grows to 30 cm in length. Often seen under jetties, this species has a pretty marbled pattern that has led to it also being called the Marblefish.

Thornfish.

Threefins

Small and easily overlooked, threefins are tiny reef fish that spend much of their time sitting on the bottom. Often mistaken for gobies or blennies, threefins have scales and a dorsal fin split into three parts, and are also called triplefins. Found in tropical and

temperate waters, Australia is home to 42 species of threefins. One of the more common east coast members of this family is the Ringscale Threefin (*Enneapterygius atrogulare*). Found throughout Queensland and New South Wales, it grows to 5 cm long. The male of this species is red with a black head, while the females are mottled brown.

Ringscale Threefin.

Triggerfish

Triggerfish have three dorsal spines, and they can trigger the first two and lock them erect to deter predators. These oval-shaped reef fish are common in tropical waters, with 22 species found in Australia. Triggerfish have a small mouth and powerful teeth,

Clown Triggerfish.

and feed on invertebrates, algae or plankton depending on the species.

Many members of this family have pretty colours, including the garish Clown Triggerfish (*Balistoides conspicillum*). This species can reach 50 cm in length and is frequently seen on the Great Barrier Reef. Another common species is the very large Titan Triggerfish (*Balistoides viridescens*), which can grow to 75 cm long.

Titan Triggerfish.

Female Titan Triggerfish are very territorial during the breeding season and guard their nest of eggs with zeal. They build a nest in coral rubble and guard a cone-shaped zone around it. Titan Triggerfish attack other fish and even divers that get too close, and many divers have ended up requiring stitches after a close encounter with one of these aggressive fish.

Leatherjackets

Australia is home to world's greatest variety of leatherjackets, with 58 species found around the nation. Leatherjackets are small-to-medium-sized reef fish found in both tropical and temperate waters. Lacking scales, leatherjackets instead have sandpaper-like skin, which has led to them also being called filefish. Leatherjackets have a dorsal spine for defence, and larger species are popular with anglers.

Leatherjacket species, Blackhead Leatherjacket (above left), Black Reef Leatherjacket (above right), Fanbelly Leatherjacket (below left) and Pygmy Leatherjacket (below right).

Most tropical leatherjackets, like the Blackhead Leatherjacket (*Pervagor melanocephalus*) are small and like to hide among corals. This species is found on the Great Barrier Reef and reaches a length of 10 cm. A widespeard small temperate species is the Pygmy Leatherjacket (*Brachaluteres jacksonianus*). This endemic species only grows to 9 cm in length and can change colour to suit its environment.

The Black Reef Leatherjacket (*Eubalichthys bucephalus*) is another endemic southern species that grows to 40 cm in length. This species is nearly always seen in pairs in areas with kelp. One of the most widespread members of this family is the Fanbelly Leatherjacket (*Monacanthus chinensis*). Found in tropical to warm temperate zones, the Fanbelly Leatherjacket has a large skin flap below the belly and grows to 40 cm long.

Cowfish and Boxfish

Box-shaped with small fins, the boxfish and cowfish are wonderful fish to watch as they slowly swim around a reef. Found in both temperate and tropical waters, Australia is home to 20 species of these fascinating fish. These fish have a hard body casing, with the cowfish also having horns on their head and body. The males, females and juveniles have very different colours, which led to many being identified as different species. Boxfish and cowfish feed on small invertebrates.

Australia's temperate waters are home to many endemic members of this family, including the Ornate Cowfish (*Aracana ornata*). This species reaches a length of 15 cm, with the males having a blue-and-yellow-checkered pattern and the females brown and cream lines. The Humpback Boxfish (*Anoplocapros lenticularis*) is another attractive southern species found off Western Australia and South Australia. Reaching a length of 20 cm, this is one member of the family where the females are prettier, with brown-and-white patterns on an orange base.

Some Australian boxfish and cowfish, clockwise from the top – Humpback Boxfish, Solor Boxfish, Yellow Boxfish (adult), Ornate Cowfish and Yellow Boxfish (juvenile).

The most abundant boxfish in tropical waters is the Yellow Boxfish (*Ostracion cubicus*). This large boxfish can reach 45 cm in length. The pretty yellow-and-polka-dot juveniles of this species are more common than the plainly patterned adults. A rarer tropical member of this family is the Solor Boxfish (*Ostracion solorensis*). Only 12 cm long, adult Solor Boxfish are blue, while the juveniles are orange and black.

Puffers

The puffers are a complex and varied family of fish, but all are poisonous and should never be eaten. Most members of this family have the ability to ingest water, puffing up to look bigger and deter predators. Puffers have prickly skin, and most eat invertebrates or algae. Mainly found on coral reefs, a limited number of puffers also reside in temperate waters. Australia is home to 57 puffers, and some are also known as toadfish and tobies.

Blackspotted Puffer.

A number of puffers get very large, with the Map Puffer (*Arothron mappa*) growing to 65 cm long. A tropical species, the Map Puffer has a wonderful skin pattern that varies as the animal grows. Many small puffers are known as tobies, like the pretty Clown Toby (*Canthigaster callisterna*). This species grows to 25 cm long, but most are much smaller. Found off the east coast, the Clown Toby is found in tropical and temperate waters. The Blackspotted Puffer (*Arothron nigropunctatus*) is a medium-sized tropical puffer that reaches 25 cm in length. Like many larger puffers it has a sagging belly, which it inflates when threatened. One of the puffers more frequently seen in temperate

Clown Toby.

Map Puffer.

waters is the Ringed Puffer (*Omegophora armilla*). Found from southern New South Wales to southern Western Australia, it grows to 25 cm long.

The Ringed Puffer is only found in temperate waters.

Porcupinefish

Closely related to the puffers, and also having poisonous flesh, the porcupinefish have long spines, large eyes and a rounded body shape. More active at night, when they feed on small invertebrates, porcupinefish can often be found sleeping during the day. These fish can also inflate when threatened, turning into a spikey ball. Most porcupinefish are tropical, but a few are also found on temperate reefs, and 12 species are found in Australia.

The Globefish (*Diodon nicthemerus*) is the most common member of this family in southern waters, but this widespread fish also ventures into the subtropics. Growing to 30 cm long, the Globefish does look like a spiky globe when it inflates. A tropical member of this family is the Blackspotted Porcupinefish (*Diodon hystrix*). One of the largest porcupinefish, this species grows to 70 cm long.

Blackspotted Porcupinefish.

Globefish.

Dense schools of Bigeye Snapper *(Lutjanus lutjanus)* are often found on the Great Barrier Reef.

THE SCHOOLERS

The following fish families are nearly always found in schools. These schools can vary in number from less than a dozen to several thousand. These fish school for a variety of reasons, but because they school they are very social creatures.

Sandtiger Sharks

Most sharks are solitary animals, only coming together to feed or mate. However, the sandtiger sharks are always found in schools that can sometimes number in the hundreds. The best known member of this family is the Grey Nurse Shark *(Carcharias taurus)*. Found off New South Wales, southern Queensland and parts of Western Australia, these impressive sharks were nearly hunted to extinction, but are now protected and making a slow comeback. Growing to a length of 3.8 m, the Grey Nurse Shark may look fierce with its dagger-like teeth, but these teeth are designed to catch fish and the shark presents no threat to swimmers or divers. Grey Nurse Sharks aggregate in schools of the same sex, and the males and females only socialise together when breeding.

Grey Nurse Shark.

Sharks reproduce in a number of ways. Some lay eggs and others give birth to live young, but the strangest reproduction strategy is used by the Grey Nurse Shark. Young Grey Nurse Sharks emerge from an egg while still in the uterus. The best developed young in each uterus then eats the other eggs and developing embryos in a bizarre form of cannibalism known as embryophagy.

Pearl Perch

Mainly found in deep water, four species of pearl perch are found around Australia. These deep-bodied fish have large eyes and are targeted by anglers. Most pearl perch gather in schools on rocky or coral reefs. Found in tropical waters, the Threadfin Pearl Perch *(Glaucosoma magnificum)* is

Threadfin Pearl Perch.

the most striking member of this family with long filaments trailing from its fins.

Growing to a length of 32 cm, this endemic species is sometimes encountered by divers off the northern coast of Western Australia.

Snapper.

Sea Bream

Almost every member of the sea bream family is popular with anglers. Ten species of these silver-coloured fish are found in Australia in both temperate and tropical waters. Sea bream feed on invertebrates and small fish, and are typically seen in schools. The largest member of this family is the Snapper *(Chrysophrys auratus)*. Reaching a length of 1.3 m, large Snapper are

Yellowfin Bream.

easily identified by the hump on their head. Normally a temperate species, Snapper also range into the subtropics. The Yellowfin Bream *(Acanthopagrus australis)* is the most common member of this species found off the east coast. Growing to 65 cm long, they are regularly seen in bays and estuaries.

Sweetlips

Possessing very large thick lips is how the sweetlips got their name. These medium-sized fish generally have pretty patterns,

Brown Sweetlips.

and most are found in small schools by day. At night they hunt over sandy bottoms in search of invertebrate species. Only found in tropical waters, Australia is home to 20 species of sweetlips.

Dotted Sweetlips.

One of the most attractive members of this family is the Oblique-banded Sweetlips (*Plectorhinchus lineatus*). Growing to a length of 50 cm, this species is found in schools on the northern Great Barrier Reef. The Dotted Sweetlips (*Plectorhinchus picus*) is a more widespread species, found from central New South Wales to central Western Australia. It grows to 50 cm long and is often found in small groups in caves. The Brown Sweetlips (*Plectorhinchus gibbosus*) is a more plainly coloured member of the family. This species reaches 75 cm in length and is found in a similar area to the Dotted Sweetlips.

Oblique-banded Sweetlips.

Snappers and Fusiliers

The snappers and fusiliers are small, colourful reef fish that are always found in schools. Some members of this family are elongated, while others are more deep-bodied like a sea bream. Found in tropical waters, Australia is home to 62 members of this family. Snappers are mostly observed swarming around coral heads, while fusiliers are constantly on the move. They feed on a variety of prey, some eating plankton, others algae, and some feed on small fish and invertebrates. Snappers and fusiliers are a popular meal for predators like trevally and barracuda, so they stay in schools for protection.

One of the most distinctive members of this family is the Red Emperor (*Lutjanus*

Neon Fusiliers.

sebae). Found from northern New South Wales to central Western Australia, Red Emperors grow to 60 cm in length and are generally seen in small schools around ledges and gutters. The Hussar (*Lutjanus adetii*) is another tropical snapper species that grows to 50 cm long. It is abundant on the southern Great Barrier Reef, where large schools gather on coral heads. Snappers often aggregate in mixed schools, especially on the Great

Yellowtail Fusiliers.

Red Emperor.

Hussar.

A mixed school of snapper and goatfish.

Barrier Reef. Two, three and sometimes four different species can be seen mixed together in schools, often with other species like goatfish.

Fusiliers are generally smaller than snappers, and most are found in schools that are constantly on the move. Neon Fusiliers (*Pterocaesio tile*) regularly sweep up and down coral reefs in their search for plankton to consume. Prevalent in tropical waters, this species grows to 25 cm long. Another common member of this family is the Yellowtail Fusilier (*Caesio cuning*), which grows to 60 cm long and has a pretty blue-and-yellow colouration.

Jewfish

Jewfish are also known as drums or croakers as they can produce a range of sounds when removed from the water. Frequently found in deep water or caves, jewfish mostly inhabit temperate waters. Australia is home to 18 species of these large elongated fish, but only one is common. The Mulloway (*Argyrosomus japonicus*) is found

Mulloway.

in schools in caves, gutters and shipwrecks from central Western Australia to southern Queensland. Mulloway are a temperate species that grow to 2 m long.

Silver Batfish

This small family of fish look very similar to bullseyes, but have more triangular-shaped bodies and prominent dorsal and anal fins. Found in dense schools in bays, estuaries and under jetties, Australia is home to three species of silver batfish. Two members of this family, the pomfreds, are found in temperate waters, but the most common and widespread species is the Silver Batfish *(Monodactylus argenteus)*. Found in tropical and subtropical waters from central Western Australia to central New South Wales, Silver Batfish grow to 25 cm long, and because of their shape are also known as Diamondfish.

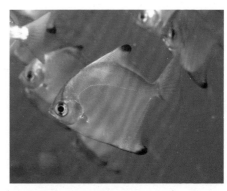
Silver Batfish.

Drummers

Drummers are oval-shaped fish that are typically seen in large schools in shallow water. These herbivorous fish feed on algae and kelp, and are found in both tropical and temperate waters. In Australia seven species of drummer are known. A tropical member of this family is the Brassy Drummer *(Kyphosus vaigiensis)*, which is often found in schools with other drummer species and grows to 50 cm long.

Brassy Drummer.

Blackfish

Found in temperate waters and feeding on algae, Australia is home to five species of blackfish. Some blackfish hide in caves and are more solitary creatures, but the most common member of this family, the Luderick *(Girella tricuspidata)*, is found in schools. A popular fish with anglers,

Luderick.

Luderick are mostly found off the east coast, but are also seen off Victoria, Tasmania and South Australia. These striped fish grow to 60 cm and are sometimes simply called Blackfish.

Moonlighter.

Stripey.

Stripeys and Mado

This small family of fish are closely related to the butterflyfish and all have pretty striped patterns. The stripeys and mado are almost exclusively found in temperate Australian waters, with five species identified. Generally found in schools on rocky reefs, this family of fish feed on plankton, algae, small invertebrates and any scraps they can find. The Stripey *(Microcanthus strigatus)* is one of the most common members of this family and found off Western Australia, New South Wales and southern Queensland. Growing to a length of 16 cm, Stripey often swarm together in ledges and gutters. The most attractive member of this family is the Moonlighter *(Tilodon sexfasciatus)*. This species doesn't school very often and is more commonly seen in pairs. Found off southern Western Australia and South Australia, the Moonlighter grows to 30 cm long.

Batfish

Also known as spadefish, because of their shape, batfish typically have a compressed circular-shaped body. The young in this family have oversized fins, and mimic dead leaves and flatworms. Batfish are generally found in groups and often form into large schools when feeding on invertebrates and zooplankton. Found in tropical

Pinnate Batfish.

111

waters, Australia is home to eight species of batfish.

Many batfish look alike, with the Tall-fin Batfish (*Platax teira*) very similar to other members of this family. This species is common in tropical and subtropical waters and grows to 60 cm long. The Tall-fin Batfish is identified by the black blotch above the ventral fins. A very distinctive member of this family is the Pinnate Batfish (*Platax pinnatus*). This species has a projecting snout and reaches a length of 35 cm.

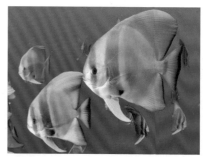
Tall-fin Batfish.

Old Wife

The Old Wife *(Enoplosus armatus)* is in a family all by itself and is endemic to southern Australia. Looking like a cross between a boarfish and butterflyfish, Old Wives have a striking black-and-white pattern and long dorsal fins. Found around kelp, seagrass and rocky reefs, Old Wives are generally found in large schools. This species reaches a length of 25 cm and feeds on small invertebrates.

Old Wife.

Sea Mullet.

Mullet

Mullets are found in a variety of marine habitats, from the open ocean to estuaries. Forming into large schools, these fish feed on zooplankton when young and algae when adults. More common in tropical waters, mullets also venture into temperate waters and some species are popular with anglers. Australia is home to 21 mullet species, with the Sea Mullet *(Mugil cephalus)* the most common and widespread. This species grows to 70 cm long and is usually found in estuaries and bays.

Parrotfish

The parrotfish are a subfamily of the wrasses with beak-like teeth. Feeding on corals and algae, parrotfish use their strong teeth to scrape off their food. Like the wrasses, parrotfish have different colour phases for males, females and juveniles, and also change sex. Only found in tropical waters, 32 species of parrotfish are found in Australia.

Bicolor Parrotfish (adult).

The largest member of this family, the Bumphead Parrotfish (*Bolbometopon muricatum*), has a very distinctive bump on its head. Growing to a length of 1.3 m, these huge fish are always seen in schools. This tropical species is often seen on the Great Barrier Reef. Many parrotfish are a greenish blue colour, like the Sixband Parrotfish (*Scarus frenatus*), which can make identification a challenge. This species is observed in small groups and grows to 47 cm long. One of the prettiest members of this family is the Bicolor Parrotfish (*Cetoscarus ocellatus*). More common in northern reef waters, this species grows to 70 cm long. Juvenile Bicolor Parrotfish only have two colours, white and orange, while the adults are multicoloured.

Bicolor Parrotfish (juvenile).

Bumphead Parrotfish.

Sixband Parrotfish.

Most sand is formed by erosion, but many of the pure white sandy beaches found in the tropics are actually parrotfish poo! When feeding, parrotfish ingest lots of coral, which is broken down into sand as they extract nutrients. This sand is then expelled as a cloud of poo. It has been estimated that large parrotfish poo almost 400 kg of sand each year.

Surgeonfish

With a sharp scalpel-like blade at the base of the tail surgeonfish should never be touched or handled. This complex family also contains the tangs, unicornfish and sawtails, and they all have tail blades and other spines for self-defence. The surgeonfish are common in tropical waters, and most have a compressed oval shape. Often seen in large schools, surgeonfish feed on algae and plankton. There are 47 species of surgeonfish in Australian waters.

Many surgeonfish have pretty colour patterns, including the Pencilled Surgeonfish (*Acanthurus dussumieri*). This species is found from central

Pencilled Surgeonfish.

Spotted Sawtail.

Spotted Unicornfish.

Western Australia to northern New South Wales and grows to 50 cm long. One of the best known members of this family is the Blue Tang (*Paracanthurus hepatus*), also known as Dory from *Finding Nemo*. Common on the Great Barrier Reef, Blue Tangs grow to 30 cm long. Unicornfish look very similar to surgeonfish, but have a spear-like projection on the head. The Spotted Unicornfish (*Naso brevirostris*) is a tropical species and often seen on the Great Barrier Reef. This species varies in colour and can change colour rapidly. The Spotted Unicornfish grows to 60 cm long. Sawtails have multiple blades at the base of the tail, with the Spotted Sawtail (*Prionurus maculatus*) having three. Found off Queensland and New South Wales, the Spotted Sawtail reaches a length of 40 cm.

Rabbitfish

With numerous venomous spines, rabbitfish are not a species to tangle with. Found in tropical waters, Australia is home to 16 members of this family. These oval-shaped fish are mostly seen in schools on coral reefs and in areas of seagrass. Rabbitfish are herbivores and feed on seaweed and algae.

Goldlined Rabbitfish.

Many members of this family have decorative skin patterns, like the Goldlined Rabbitfish (*Siganus lineatus*). This widespread species is found from southern Queensland to central Western Australia and grows to 47 cm long. Another attractive member of this family is the Masked Rabbitfish (*Siganus puellus*). Always seen in pairs, this species grows to 38 cm long and has a distinctive black mask.

Masked Rabbitfish.

The Silvertip Shark *(Carcharhinus albimarginatus)* is a member of
the whaler shark family and mostly seen on reefs in deep water.

THE PELAGIC WANDERERS

Most fish live close to the bottom, as this is where they find shelter and food. However, a number of fishes have taken to the open seas, living a life as pelagic wanderers. Finding food in the open ocean is not easy, so many of these pelagic wanderers travel vast distances in search of a meal.

Mackerel Sharks

The most notorious of all the sharks, the Great White Shark *(Carcharodon carcharias)* is a member of this small shark family. The mackerel sharks are medium-to-large sharks that feed on fish, rays, sharks and marine mammals, depending on the species. Mostly found in temperate waters, mackerel sharks are warm-blooded animals that need a lot of food to maintain their elevated body temperature. Australia is home to four species of mackerel sharks, but only the Great White Shark is considered common. These impressive sharks grow to 6.5 m long and have triangular-shaped teeth to saw through flesh. They are found around Australia, but have a preference for cooler waters.

Great White Shark.

Whale Shark

The Whale Shark *(Rhincodon typus)* is the sole member of this family and the world's largest fish. These huge sharks grow to 14 m long, but have tiny teeth as they feed on plankton. Found in tropical and subtropical waters, the only place where they are known to aggregate in Australia is

Whale Shark.

Ningaloo Reef, Western Australia. Whale sharks gather in this area from March to August to feed on the plankton soup associated with the annual coral spawn. Local dive operators take snorkellers out to encounter these gentle giants.

Whaler Sharks

The whaler shark family is very diverse, with 31 species found around Australia. Whaler sharks are medium-to-large sharks, and some members of this family are

Bull Shark.

Grey Reef Shark.

Whitetip Reef Shark.

potentially dangerous. Most members of this family live in tropical waters, but a few venture into temperate zones and some also inhabit freshwater rivers. The whaler sharks are very efficient hunters, feeding on fish, rays and other sharks.

One of the most common members of this family, found on coral reefs, is the Whitetip Reef Shark (*Triaenodon obesus*). This species is quite timid and considered harmless, and grows to 2.1 m long. Often observed patrolling coral reefs, the Whitetip Reef Shark is one of the only members of this family that rests on the bottom by day. Another common reef species is the Grey Reef Shark (*Carcharhinus amblyrhynchos*). The most dangerous whaler shark is the Bull Shark (*Carcharhinus leucas*). This very solid shark grows to 3.5 m long and is found in tropical to warm temperate zones. Able to move from salt to freshwater, Bull Sharks often feed in murky rivers and canals, which has seen them accidently bite swimmers.

Eagle Rays

Eagle rays are bottom feeders, consuming molluscs and crustaceans, but spend much of their time soaring through the water column. These graceful rays love cruising in currents and are often found in schools. Eagle rays have a protruding head, a long tail and a tail spine, but are not considered dangerous. Found in both tropical and

Spotted Eagle Ray.

temperate waters, Australia is home to five species of eagle rays. The most common and widespread member of this family is the Spotted Eagle Ray *(Aetobatus ocellatus)*. Found in tropical and subtropical waters, the Spotted Eagle Ray can have a wingspan of more than 3 m, but rays that large are very rare.

Reef Manta Ray.

Devil Rays

The devil ray family contains the manta rays and mobula rays. These giant rays have a large mouth flanked by cephalic fins that funnel plankton for swallowing. The mobula rays have a narrow head and tail spines, and are generally shyer and smaller than the manta rays. Both types wander the oceans, only visiting reefs to feed and get cleaned. In Australian waters are five species of devil rays, with the Reef Manta Ray *(Mobula alfredi)* the most common species. These majestic rays grow to 5.5 m wide and are found in tropical and subtropical waters.

> Reef Manta Rays are very curious creatures and often investigate divers and snorkellers. They have a large brain, compared to other sharks and rays, and are considered to be the most intelligent of all the fishes.

Suckerfish

With a modified dorsal fin that acts like a suction cap, the suckerfish have adapted to a life attached to other marine life. These strange elongated fish may appear to be nothing more than a large parasite,

The Slender Suckerfish is also called the Remora.

but they do perform cleaning duties on their host. Suckerfish attach to sharks, rays, turtles, gropers, whales and even boats, and generally feed on scraps missed by their host, and their host's poo. Mainly found in tropical waters, Australia is home to seven species of suckerfish. The Slender Suckerfish *(Echeneis naucrates)* is the most common member of this family. These striped fish grow to 1 m in length and sometimes a dozen or more crowd onto a host.

Trevallies

The trevallies vary greatly in shape and size, with this family also containing the kingfish, scad and darts. Trevallies are found in both tropical and temperate seas, and have elongated bodies with a silver metallic sheen. Most species are pelagic, moving

School of Bigeye Trevally.

from reef to reef or wandering the open ocean. Trevallies are generally found in schools and all species are carnivorous, feeding on smaller fishes. Trevallies are very common around Australia, with 67 species identified.

One of the larger members of this family is the Yellowtail Kingfish (*Seriola lalandi*). Common in temperate waters, these bullet-shaped fish grow to 2.5 m long and are popular with anglers.

The Rainbow Runner (*Elagatis bipinnulata*) is a tropical member of this family and typically found in large schools. These pretty fish have two distinctive blue lines and grow to 1.2 m long. The scad are the smallest members of the trevally family and often seen on reefs or under jetties. The Southern Yellowtail Scad (*Trachurus novaezelandiae*) is found in temperate waters from central Western Australia to southern Queensland. This species can grow to 50 cm long and is hunted by other members of the trevally family.

Many trevallies are seen in tropical waters, with Bigeye Trevally (*Caranx sexfasciatus*) very common. This species is always found in large schools, swarming around coral heads or under boats. Bigeye Trevally grow to 85 cm long and are identified by the white tips on their fins. A very large member of this family is the Giant Trevally (*Caranx ignobilis*). Found in tropical and subtropical waters, Giant Trevally can reach 1.7 m long and are generally found in pairs or small groups. The darts are another member of the trevally family, with elongated dorsal and anal fins. The most distinctive member of this family is the Snubnose Dart (*Trachinotus blochii*). Found in tropical waters, this species has a rounded head and grows to 65 cm long.

Giant Trevally.

Rainbow Runner.

Snubnose Dart.

Yellowtail Scad.

Yellowtail Kingfish.

Barracudas

Barracuda have a reputation as being very dangerous, but in reality these fish only snap at anglers that catch them. Barracuda have elongated bodies and feed on other small fishes, which they catch with their large teeth. Found in tropical and subtropical waters, most barracuda are found in schools soaring in midwater. Australia is home to ten species of barracudas.

Blackfin Barracuda.

The largest member of this family is the Great Barracuda (*Sphyraena barracuda*). These impressive fish grow to 2 m long and are more solitary than other species. Many species of barracuda have rows of dark bars

Great Barracuda.

along their body, making identification difficult. The Blackfin Barracuda (*Sphyraena qenie*) has twenty of these dark bars and grows to 1.7 m long. This species is often seen on the Great Barrier Reef.

Mackerels and Tunas

Members of this family are powerful ocean wanderers, and also an important source of food in many cultures. The mackerels and tunas are members of the Scombridae family, which also includes the wahoos and bonitos. These impressive fish are streamlined and very fast through the water, preying on smaller fish, invertebrates and cephalopods. The tuna are also very unique in having warm blood, which helps to make them fast and efficient swimmers.

Shark Mackerel.

123

Many of these fish make large migrations, and are found in both tropical and temperate seas. Australia is home to 25 members of this family.

A wide-ranging member of this family is the Shark Mackerel (*Grammatorcynus bicarinatus*). More common in tropical waters, this species also ventures into the temperate waters of New

One of the most prized fish in temperate waters is the Southern Bluefin Tuna.

South Wales and Western Australia. The Shark Mackerel looks very similar to other members of the mackerel family and can be identified by the small black spots on its belly. This species grows to 1.1 m and is generally a solitary animal.

The most impressive member of this family is the Southern Bluefin Tuna (*Thunnus maccoyii*). These massive fish grow to 2.3 m long and can live up to 40 years. Being a popular food source has seen this species depleted in numbers. Southern Bluefin Tuna are found off southern Australia, but migrate into warmer waters to spawn.

Sunfish

Known for their habit of basking on the surface, sunfish are one of the oddest families of fish found in Australian waters. These large disk-like fish look like they have been cut in half, with oversized dorsal and ventral fins and a short stumpy tail. Sunfish are the heaviest off all the bony fishes, weighing up to 2300 kg, and feed on a wide range of creatures they encounter in open waters, including sea jellies, small fish and plankton. Found in both tropical and temperate waters, Australia is home to five members of the sunfish family, including the enormous Bumphead Sunfish *(Mola alexandrini)*. These enormous fish can grow to 3 m long and 4.2 m in height. Cruising the open ocean, encounters with Ocean Sunfish are very rare, but they are known to visit shallow cleaning stations at times.

Bumphead Sunfish.

Other Natural History titles by Reed New Holland include:

Australian Marine Life: The Plants and Animals of Temperate Waters
Graham Edgar ISBN 978 1 87706 948 2

Australian Tropical Marine Wildlife
Graham Edgar ISBN 978 1 92151 758 7

Coral Wonderland: The Best Dive Sites of the Great Barrier Reef
Nigel Marsh ISBN 978 1 87706 7808

Deadly Oceans
Nick and Caroline Robertson-Brown ISBN 978 1 92151 782 2

Diving With Sharks
Nigel Marsh and Andy Murch ISBN 978 1 92554 600 2

Field Guide to the Crustaceans of Australian Waters
Diana Jones and Gary Morgan ISBN 978 1 87633 482 6

Fishes of Australia's Southern Coast
Martin Gomon, Diane Bray and Rudie H. Kuiter ISBN 978 1 87706 918 5

Guide to Sea Fishes of Australia
Rudie H. Kuiter ISBN 978 1 86436 091 2

Muck Diving
Nigel Marsh ISBN 978 1 92151 781 5

Seabirds of the World
David Tipling ISBN 978 1 92151 767 9

Tropical Marine Fishes of Australia
Rick Stuart-Smith, Graham Edgar, Andrew Green and Ian Shaw ISBN 978 1 92151 761 7

Underwater Australia
Nigel Marsh ISBN 978 1 92151 792 1

Wild Dives
Nick and Caroline Robertson-Brown ISBN 978 1 92554 642 2

World's Best Wildlife Dive Sites
Nick and Caroline Robertson-Brown ISBN 978 1 92151 772 3

For details of these and hundreds of other Natural History titles see
newhollandpublishers.com

And follow Reed New Holland and New Holland Publishers on Facebook and Instagram

Index